THE OTHER SIDE
OF THE DESK

A story about a chronic pain specialist
who became a chronic pain patient
& his advice for chronic pain sufferers

Stuart Donaldson, PhD

www.isnr-researchfoundation.org

For information about this book:

ISNR Research Foundation
1925 Francisco Blvd. E. Ste. 12
San Rafael, CA 94901
(415) 485-1344
www.isnr-researchfoundation.org

Cover Design and Layout
Cynthia Kerson, PhD

10-digit ISBN 0984608559
13-digit ISBN 978-0-9846085-5-3

2012 ISNR Research Foundation
executivedirector@isnr-researchfoundation.org

All images are free domain

"The Other Side of The Desk: A story about a chronic pain specialist who became a chronic pain patient & his advice for chronic pain sufferers" is a publication of the ISNR Research Foundation. Opinions expressed herein are those of the author and do not necessarily reflect the official view of the ISNR Research Foundation.

Printed in the United States of America

Contact the author:
Dr. Stuart Donaldson
myo@cubedmail.com

TABLE OF CONTENTS

ACKNOWLEDGEMENTS

Special Acknowledgements:

- To Dr. Stephen Miller who put an ugly pelvis back together: a brilliant orthopaedic surgeon and a humanitarian.
- To Dr. Robert (Dr. Bob) Schulz, Professor University of Calgary who for years has supported and helped push forward the ideas presented in this book.

Thanks:

- To the over 1,000 patients we have seen over the years.
- To the current staff (Leslie, Doneen, Mary, Pattie, Marly, Donna and Shirl) and all the previous staff at the clinic.
- To the numerous therapists who worked with me over the years (Ellie, Pattie, Donna, Tamsin, Charlene, Rick, Channone, Bob, and Julie)
- To Arlene Devrome medical/psychological confident
- To Lawrence Klein, Bob Clasby, Jill Brash, Dr. Horst Mueller, Dr. Henry Svec, Rick Macnab and all the others from AAPB who listened over the years
- To Dr. David Romney, University of Calgary
- To Drs. Emerson Gingrich, Noel Purkin, and Simon James who cared for me
- To Mike Harrison who edited this book and first gave me the idea to write it.
- To Peg Ainsley for improving the manuscript's readability and making it publisher-ready.
- And finally to Dr. Gabe Sella who taught me much about physiology, the medical system and wrote numerous research articles with me.

FOREWORD

UNDERSTANDING PAIN AND ESPECIALLY CHRONIC PAIN is a work in progress. While pain practitioners may view new concepts and treatments with well-guarded optimism, the chronic pain sufferers are more than eager to adopt any new method of treatment that enhances the hope of healing and redemption.

This book finds its focus at the interface between the well-intended long-term view of scientific and clinical endeavor to diagnose and treat chronic pain in a Hippocratic manner and the need of the chronic pain sufferer to get rid of the scourge in the least time-frame possible.

It is rare to read a book where the author enables the reader to identify with two personae: the clinician and the patient. This is such a book. Who will benefit from reading and studying it? If I have a choice of one word, that word would be "everybody."

It is hard to find a person in our society who is not aware of or acquainted with someone in the family or social circle who is not a pain sufferer.

This book will be of special benefit to all clinicians from any field that deals with pain or any aspect of it. It will greatly enhance the understanding of chronic pain of students and masters alike.

How about the pain sufferer? Who would not appreciate sharing the understanding of ongoing, relentless pain with a clinician and pain sufferer alike? Who would not appreciate the reading of so many facts that extinguish the shame one feels when one hears "it's all in your head?"

Furthermore, it is a rare recognition of the role of the ancillary pain office personnel. All those gate keepers who are not even spoken of in the clinical lectures and are not recognized for their true value can find an ally in this book. The pain patients may gain new insight in the way they interact with the clinician's office personnel and with the technical and clinical personnel.

Lastly, the book may serve as an eye opener to all those clinicians who learned to belittle pain by calling it "discomfort" (unless it happens to them or their families). It may enhance the hope that they will become more humane and humble in their approach and teaching of a new generation.

To end on a personal note, I have known Dr. Stu Donaldson and his exceptional life partner, Mary, for almost twenty years. I know them as scientists, clinicians and especially as friends. Only someone in a position such as mine could appreciate the many truths stated by the author with such parsimony of words and such power of emotions.

Gabriel E. Sella, MD, MPH, MSc. PhD (HC)   July 2010

INTRODUCTION FROM THE AUTHOR

IT HAS BEEN THREE YEARS SINCE I started writing this book, which has been quite an adventure. I started writing it at the encouragement of a friend who thought I had an incredible story to tell: a story of a specialist in pain management who became a chronic pain patient. At first I was simply caught up in the idea of writing, it was good to get my brain challenged again. Then I became cathartic as experienced the emergence of anger, despair and hope, mixed with the deeper understanding of pain. So I wrote the book as a dialogue between Dr. D the pain specialist and Stu the patient. That edition was promptly rejected by a publisher as too difficult to read because the shifting between the two personas was confusing. Write the book for the general public not the academics came back the feedback.

The second rewrite feedback came back that it was not academic enough.

This difference in opinion between the reviewers illustrates to me the problem between the pain professional and the patient: it is a vast schism. Neither one is right, but it is imperative that each understands the other. It is my hope that this book bridges that schism, leading to improved health, reduced pain and better quality of life.

Stuart Donaldson   June 2012

PREFACE

The difference between acute pain and chronic pain has nothing to do with the severity of the pain, merely the longevity. When you first see your family doctor about a specific pain, he will attempt to diagnose the cause and will probably send you for tests and prescribe pain medication. If the pain persists, your family doctor will probably send you to see a specialist. Usually by the time you see the specialist, your pain is no longer acute; it has now become chronic. The time period during which acute pain becomes chronic pain is usually around six months. Sadly, if the pain or its cause(s) aren't life threatening you might now remain in chronic pain for many, many years.

At some point, this chronic pain might motivate you to seek alternative medical advice. This is why many people end up in Dr. Stuart Donaldson's office. Dr. Donaldson is a clinical psychologist who specializes in treating chronic pain patients. He is a recognized world leader in a wide variety of clinical techniques, including the field of biofeedback and the newer science of neurofeedback, and he has helped a growing number of chronic pain patients to reduce and manage their pain more effectively, thereby reducing their dependence on prescription medication and other forms of treatment.

The majority of his patients have experienced a sports injury or a car accident, or maybe an overly ambitious attempt to exert themselves, resulting in the onset of pain. These are the "there but for the grace of God go I" moments that come as such a surprise when it happens to us. It certainly came as a surprise to Dr. Donaldson the day he suffered a similar fate after he slipped on an unseen piece of ice. It was this defining moment that changed his life forever. He went from the doctor who helped alleviate pain to the patient who needed all the help he could get.

As his injury went from acute to chronic, he continued to believe that he could use his expertise to make the pain go away. Little did he know his pain would get worse and would continue to plague him for the next ten years. Despite his training and vast knowledge of pain management, Dr. Donaldson was almost powerless to manage his own, worsening chronic pain.

Dr. Donaldson's encouraging story is one of anguish, family upheaval, distress and disappointment. He tells his readers how the years of chronic pain threatened everything he held sacred in his life. And yet, ultimately, it is a story of hope because, after almost ten years of constant pain, he has managed to come to terms with it and his journey through these trying times offers renewed hope to us all.

This, then, is Dr. Stuart Donaldson's remarkable journey.

CHAPTER ONE

My Accident

~~~~~~~

IT'S NEW YEAR'S EVE 2000.

I was taking Mary, my wife of twenty-six years, out for supper at our favourite restaurant. What better way to celebrate the arrival of another new year? The weather had been beautiful all day, clear blue skies and lots of sunshine. Melted snow ran from the rooftops, running in streams through the downspouts, out across the sidewalks and down the storm drains all across town.

But in the shadows lay a danger that was to be my undoing.

Mary and I were both looking forward to a relaxing supper, just the two of us. I parked close to the restaurant and shut the engine off. Mary looked lovely.

We both stepped out of the car. I walked quickly around to Mary's side, intending to take her arm before we headed for the restaurant. I made it as far as the sidewalk and my foot struck a patch of ice.

I went down hard, like a proverbial sack of potatoes. It surprised the heck out of me. One minute I was upright, the next I was on my butt, gasping for air, a strange sensation fogging my brain. I'd fallen squarely on my hips and, though I could feel the cold, wet pavement soaking through my suit pants, I wasn't aware of any other sensation. I was too confused to stand up. The fall had knocked the wind out of me and I was dazed and bewildered from the impact. It felt as though I'd been hit by a freight train and then run over by an eighteen-wheeler.

I remember Mary, with a look of great concern, asking if I was all right. At first, I felt too stunned to answer. And what was I going to say? I was struggling to understand why my mind was in such a state, so cloudy and disoriented. I had been a trainer on a junior hockey team for eight years and I was a licensed EMT-A - trained in crisis intervention. After what seemed like an eternity my years of practice kicked in. I wiggled my toes, my ankles and my knees. Everything seemed to be moving just fine. I looked up at Mary and assured her I was okay. Yet this assessment was performed by remote control; it was as though I was on automatic pilot. I had been trained to go cold and clinical in an emergency and that's exactly what I did. Mary helped me to stand up and I began to put weight on my right leg.

So far, so good.

Then it happened.

As I put all my weight on my right side, there was a loud *snap* from somewhere deep inside my leg.

Then nothing.

No pain, no abnormal sensations, just a single, sharp *snap*. My mind was still clouded by fog but I managed to spread my weight between both legs. I rested a moment, and then proceeded, stubbornly, to the restaurant, ignoring all the danger signs like the jackass I know I can be. Somewhere in the very back of my mind I knew I should have allowed Mary to call an ambulance. But pride, as the saying goes, comes before a fall.

In my case, it came afterwards.

I don't remember what I ordered for supper that night though I do remember I didn't eat very much of it. I know I downed a few double whiskies, not for the pleasure, nor for the pain; I drank the whisky for the shock. And for the numbness, disorientation and confusion that had taken over my brain. I thought the whisky would somehow kick-start me back to life.

This might be tough to understand to one who has never been through serious trauma. Of course, at the time, I had no idea of the seriousness of my fall. All I knew was that something had gone awry and for the next hour and a half I struggled through supper in a daze. According to Mary, I was able to carry a conversation, although I have no memory of it. All the while my head felt as though it was in a thick pea soup, a dense fog. Yet I was perfectly able to function and would have scored magnificently on the Glasgow Coma Scale. I was awake and alert but just not connected to anything of this earth.

I realize now that my loss of concentration and inability to focus were symptoms of concussion. When I see patients with this complex I now suspect concussion as part of their injury. I am also looking for signs of ongoing trauma including tremors and anxiety, which are created when the force of impact is transmitted up the spine potentially impacting the brain stem.

Then the pain arrived. It began as soreness in my hips, rising rapidly to an uncomfortable level. We cut the dinner short and Mary drove us home. I think the sensation of pain was my brain's way of cutting through the fog, prodding my sensory system back to life. I took some over-the-counter painkillers and climbed slowly into bed. I slept well that first night but woke up to bright, crisp pain in the early morning.

The fog had lifted. There was nothing to stop the pain from slicing through me. The painkillers had worn off and the pain in my hips was front and centre. Both hips were killing me, especially the right one. I was angry with myself, feeling stupid that I'd fallen and that the injury was going to mess up my plans for the next few weeks. As I lay in pain, failing to find a comfortable position, I realized I needed to see my doctor and arrange for an x-ray to rule out any serious damage. My self-assessment assured me that nothing was broken, since I'd made it through supper and had managed to get myself home and upstairs to bed.

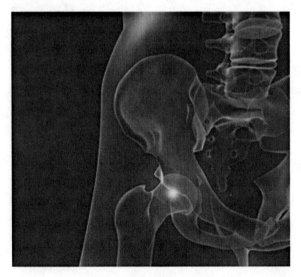

As the day wore on, I wore out. The pain in my hips got progressively worse and a deep ache set in. It was very uncomfortable and I retired back to bed with a strange combination of ice packs and hot water bottles. The ice seemed to help the parts that were tender and swollen while the parts that were painful and aching responded better to the heat.

The following day, I managed to visit my doctor who arranged for a series of x-rays, taken from the front towards the back.

The results came back negative, just as I expected; no surprise there. They just showed some muscle bruising and contusions.

Two weeks and I should be as good as new, said everyone I consulted. You'll be back to your cantankerous old self by the middle of January.

But that's not what happened.

IF YOU ARE GOING THROUGH HELL, KEEP GOING.

~ WINSTON CHURCHILL ~

CHAPTER TWO

Background

~~~~~~~

CHRISTOPHER CHARLES STUART DONALDSON WAS BORN in Lethbridge, Alberta on December 3, 1945, son of Adam (Addie) Donaldson and Margret (Peggy) Donaldson with an older sister Diane.

Our family was the customary family of the times; immediately post WWII; Dad was the wage earner, mum was the stay-at-home mother. Dad was the manager of a coalmine with 2,700 men working for him. However, Mum ruled the house. I remember being taught, "Don't argue with her, she is your mother." My family life was one of happiness and doing things together.

Sports played an important role in our family as my dad was not only a good athlete, but believed in the value of sports as a means to develop team work and keep young men off the streets. Dad had been recognized for these achievements by being inducted into the Lethbridge and Alberta Sports Hall of Fame, for which I am very proud. This is the essence of my family's culture.

My preschool years were plagued with medical issues beginning at six weeks when I sustained a broken collarbone. With no apparent reason for this; I just started crying and crying. The doctors finally figured out what was wrong. I managed to achieve five more breaks in both collarbones before the age of six. Eventually it was determined I had what is termed "brittle bones." I also developed pneumonia four times in that six-year span.

Today, well-meaning people would immediately suspect that I was being abused, although that was not the case. I recall being raised in a household of love and caring with rules that were generally enforced including being spanked on the odd occasion.

Despite the medical issues, I enjoyed life, was part of a small community, had many close friends and did well in school. I easily attained passing grades and it was expected that I would go to university and become a doctor or engineer. Until age eight, life was good and comfortable.

Yet another medical menace entered the picture at age eight: polio! In 1953 an epidemic was sweeping North America, killing and paralyzing thousands. I had not been feeling well so mum took me to the doctor. I was fairly accustomed to doctors by now and I waited unconcerned. Next thing you know they are jabbing me in the spine with this monstrous painful needle. After the screaming, yelling and crying subsided I was whisked off to a special hospital. Not just any hospital, but one that only took polio patients.

In those days one of the ways of combating the spread of this disease was to isolate the patient, so into isolation I went. In just two hours my life had been turned upside down. I was placed in a bed, frightened - not knowing if I was going to live or die with no family contact or support. A guy in the bed next to me was desperately trying to breathe. Lying there I could hear gurgling sounds as the air moved in and out of his fluid-filled lungs. A nurse brought me something to make me sleep. In the morning I awoke alone in a quiet, empty room with no bed or person beside me.

In 1953, except for the medical staff, isolation meant complete isolation, as even my parents could not come to see me. Mum would come every day at 2PM and stand on the sidewalk outside and wave to me for half an hour. She was my only contact with the outside world. To this day I still get choked up remembering the overwhelming fear and loneliness I felt and how I wanted to be with my mother. After two weeks the fever from the polio broke and joyously I left the hospital with some residual paralysis on my right side especially in the leg, but damn glad to be out of there.

While recovering at home, a pain started to develop in my right hip. Initially thought to be a residual effect from the polio, it turned out to be Leg Perthes Disease. "This is not so bad," I thought, as I got to stay in bed for two weeks missing school. Mum seemed really upset but I was fine compared to what I had just been through. Two weeks later I was back to normal, with no apparent ill effects. The full implications of this disease would only become apparent much later in life.

Although my father still held out hope that I would do well in sports, it became clear that I was simply unable to participate at any sort of competitive level. While I strove to please my father it was frustrating that I just was not physically able to do so.

I did not grow to be very tall or strong but I didn't have any real pain or deficits except that I walked with a limp. My physical activity consisted of gardening, building things about the yard, golfing, curling and being generally physically active. I did not feel physically handicapped, nor did I feel much pain except for occasional aches in the right hip.

After all my experiences with the medical profession I decided that I wanted to be a doctor for they were my heroes. I didn't have an area of interest in mind or a plan in place but I sure felt I had the smarts to succeed.   In 1953, when I was 13, the coalmine shut down and dad got involved in the oil patch. We moved to Calgary where I settled into Henry Wise Wood High School, graduating in 1963 majoring in the 3 Bs: booze, broads and bridge. I was continually in trouble with the school counsellor and I really did not strive to "achieve to my potential." Maintaining a healthy position on the laggards list, I would take delight in frustrating the heck out of the guidance counsellor by just barely passing. I had figured out that I had a brain and with a minimum of work I could pass. University entrance required a 60% average on grade 12 final exams. I got 61% much to his disgust.

Attendance at the University of Calgary came next, and I continued to major in the 3 Bs. This time my poor work habits caught up to me. This plus another major medical problem, kidney disease, created much absenteeism and got me expelled for a year due to poor grades. To add insult to injury, the girl I had dated for over a year, my first true love, dropped me on my head.

I hit rock bottom. Things had always worked out for me, until now. Lost, confused and totally depressed I moped around the house for a month not accomplishing anything except for lots of walks with the dog. I still received unconditional love from my parents, each in their own way. Mum supported me and offered encouragement. Dad kicked me in the butt and told me to get off my ass and get going, reminding me that nobody was going to do it for me, that it was up to me.

It was expected that if I was not in university that I would work. Most jobs requiring physical size or muscle strength were out of the question. Finally I got a job holding an elevation rod as part of a crew running a survey line near the Alberta/Saskatchewan border. It felt like I walked a thousand miles in the fall and winter of 1964 and into January of 1965. One day in January I learned a new definition of cold.

The temperature dropped to minus 40 degrees Centigrade with a 40 mile an hour wind. It was so bad that through a heavy-duty ski jacket I got frostbite on the back of my neck. My neck is cold sensitive to this day. We finally took refuge in a small town to wait out the storm. The survey crew decided to go drinking and because I was underage I sat in my room staring out the window. I was overcome with the eerily similar feeling I had looking out the window of the isolation hospital years before. This time there was no mother to stand and wave to me and I realized then and there I had to grow up and get on with my life. I decided that university was easier than this life and it was time to plan to move ahead.

Going back to school was tougher than I expected and I just barely got through undergraduate studies at the University of Alberta. I took mainly psychology, sociology and other humanity courses and still struggled through for I did not really know how to study. I dropped math, as there was no chance of passing it, and replaced it with philosophy during summer school.

Psychology became my major, with a minor in sociology. After the first year my marks started to improve and my interest in psychology grew. However I also continued my love of the 3 Bs and was pleased to join Phi Kappa Pi fraternity. In my growing up years our family spent a lot of time in the Legion Hall in Lethbridge, as this was the place for the war vets and the miners to hangout. I felt comfortable in the fraternity since it reminded me of the atmosphere in the Legion.

Gradually I was getting to understand what it took to succeed and yet have fun. My life was a composite of studying, fraternity activities and dating, and I met a young lady from Quebec in my last year of undergraduate studies. After graduation in 1968 I planned to head off to Toronto in pursuit of this young lady. Before leaving for Toronto, I worked for six months as a psychological assistant with Alberta Mental Health Services in Grande Prairie, Northern Alberta. It was a time of seeing life from the other side of the tracks, treating depression, marital problems, drug addiction and the like. It was a totally different reality from what I was used to in terms of family values and family care.

When I did move to Toronto, I worked in a crisis intervention program and trained in numerous medical and psychological techniques, including psychiatry, paediatrics, general medicine and neurology as it applied to psychology. This was the best thing that ever happened to me; I started to really understand the human body. I was especially interested in neurology. This focus and direction started to evolve my professional life, stimulating my brain like nothing before it. The 3 Bs were still around, but now I was shifting gears and devoting time and energy to this new area of learning.

I was excited to realize that I had a good brain and could move forward in this direction. I worked in a crisis unit that mainly dealt with suicide cases, which matured me and put a strain on me emotionally.

I was both stunned and angry when a patient killed himself.. Stunned because I had just seen this patient that morning. Angry because I had not said anything in the pre-discharge meeting even though I had some concerns about him. Three more patients from the unit would kill themselves that year, which was never easy to accept. This certainly made me take a look at my life and what I was doing with it. I resolved to use my strengths to help people in any way I could, but also to never feel responsible for somebody else.

I returned in 1973 to the Calgary area. In early 1974, while I was working with Alberta Mental Health Services in a small town south of Calgary, I met my future wife, Mary, and proposed six weeks later. A strong intellectually gifted person, independent and quietly warm hearted I just knew she was the one for me. It was time to settle down and focus on a career. In 1976 I went to graduate school at the University of Calgary and graduated in 1977 with a Master's degree in science. I had finally learned to use my abilities to reach my goals.

In the early 1980s two seemingly small events occurred that profoundly impacted the subsequent years. One was a conversation that Mary and I had with my father, the other a conference I attended focusing on biofeedback.

One day when Mary and I were visiting Mum and Dad, Mary asked Dad the question "Why did he like sports so much?" Mary's background had been much more arts- and literature-based and she just did not understand the fascination with sports. My dad's answer I will never forget. You see, he said "Mary, as we go throughout life we will never know if what we do today is successful; if we won; for there is always tomorrow and another problem and another challenge. Sports are the only thing that at the end of a pre-determined time period we declare a winner. Someone we can clearly say they won and are the best at that moment. This never happens in life for there is no time at which the clock stops except in death." My dad passed away shortly thereafter but his words still stick with me, for in dealing with chronic pain while we can never say it is cured we now have techniques that allow us to win a day at a time.

The other event was my stumbling upon the field that was to shape my life's work when I attended a weekend conference on biofeedback hosted by the Psychology Department at the University of Alberta. Imagine having your eyes opened to a new and exciting field that combined all of your training with what you had been taught as a youngster, and new and innovative ideas on how to treat physical problems. I found my chance to achieve the goal of helping people in pain. I came back excited, ready to take on new learning and challenges.

I saw the opportunity to help and heal that I had imagined as a child. No more id, ego, and super ego, just data to direct treatment. This conference directly led to the establishment of our clinic and my move into private practise. I had arrived to an exciting new place in my life and in my mindset.

This excitement propelled me to go back for my PhD focusing on chronic pain, pain management and biofeedback. Excited and challenged with this education I focused on chronic low back pain and its treatment with a biofeedback technique called surface electromyography (SEMG). I graduated from the University of Calgary in November 1989 with a PhD in Clinical Psychology, moving quickly into private practice.

With the support of my wife, Mary, my staff and my colleagues, the clinic started to flourish. Initially our focus was on low back pain, resolving this issue with a great deal of success. We named the clinic Myosymmetries, which is Latin for 'muscle balance' and our focus was to make the muscle activity on either side of the spine equal in strength and activity using SEMG techniques. With this success our reputation grew and we were soon referred to for other types of pain such as carpal tunnel syndrome and headaches. We were privileged to work with some Olympic athletes. The clinic got involved in research and publishing papers in our ever-expanding areas of interest and we traveled about the world presenting our findings.

Then came the accident.

A LIFE SPENT MAKING MISTAKES IS NOT ONLY MORE HONORABLE,
BUT MORE USEFUL THAN A LIFE SPENT DOING NOTHING.
~ GEORGE BERNARD SHAW ~

CHAPTER THREE

After the Accident

~~~~~~~

MY EXPECTATION WAS THAT I WOULD RECOVER and life would return to the way it was before the accident. The people we've seen at the clinic over the years all voiced the same expectation or desire: I want my life back to the way it was before the accident.

I had been taught that the 'normal" healing time for breaks, strains and sprains is about six weeks. It is expected that the pain from these types of injuries will decrease during this time period. However, myofascial pain (pain originating in muscles or surrounding fascia) appears after about six weeks becoming more intense over time and with increased activity.

My pain did not follow this pattern. During the weeks immediately after the fall, my pain increased. The pain became more intense or severe especially in the right hip, gradually reducing my ability to bear weight, walk or stand. Crutches went from being an occasional necessity to being a full time requirement. This was not only confusing but it was starting to make me angry. It did not appear to be myofascial pain. The damn pain was interfering with my work. It required the staff to compensate for me.

The continuation and worsening of the pain was inconsistent with the X-ray results, which indicated there were no breaks. If it was only muscle pain there should be a period post-trauma as described; there should not have been so much pain, not pain immediately after and continuing to get worse!

With some hesitation and after a painful four weeks, with the advice of my doctor, I began physical therapy. Being a professional in the healthcare field I knew the various therapies and therapists and selected ones who I considered to be the best. During the first session, after a series of exercises, which really did nothing for the pain, Interferential Therapy, an electrical pain blocking technique, was applied. Interferential Therapy uses a mid-frequency electrical signal to treat muscular spasms and strains. The current produces a massaging effect and stimulates the secretion of endorphins to relieve the pain. Strained muscles relax and promote soft-tissue healing.

Pain shot through my groin, buttocks and right leg the moment they turned the current on. I was in agony, almost screaming. After 15 minutes of undiminished pain, I told them to call an ambulance. I wound up in an emergency department after a ride I don't remember because of the nitrous oxide (laughing gas).

They did not do anything for me in emergency. They relied upon my X-rays, and since they were negative, they gave me a tranquillizer. I am sure they thought I wanted drugs, diagnosed me as hysterical and sent me home with a prescription for a tranquillizer, which I threw out.

What a cocktail of emotions: I was angry, how dare they think I was a drug addict, I was sincerely troubled and puzzled and I was scared. I had never experienced pain like that. Where was it coming from? I had no broken bones. Fear that I had a rare form of something, perhaps cancer, was building. The pain was not consistent with what would be expected from myofascial problems! I was trying to keep control by distancing myself from the pain through intellectualizing everything; trying to be logical in this sea of emotions.

For a month the pain was getting worse, requiring almost constant ambulation on crutches. Anger started to dominate my emotions. I wondered if there is a physical problem that has been missed.

At the clinic I am continuing to function, albeit not well, as the pain keeps causing me to lose my focus and concentration. As a therapist, I felt responsible to my clients and did not take painkillers during the day. I worry about seeing a patient while stoned. However, it is tough to focus when the right hip is throbbing and you are tired and emotionally drained by pain.

Pain demands your attention, despite what the experts say about using distraction techniques to deal with it. All I can say is try distracting yourself 24 hours a day, seven days a week. I could not do it. Distraction techniques only worked while doing them. I could not do them all day every day. Besides I was too damn tired.

Finally enough pain. Exhausted and angry, embarrassed and upset, I think "Can't cure himself, how is he going to help others?" I decide an MRI is needed and refer myself for one.

As I enter the MRI center I feel an assortment of emotions. I don't know why but I feel embarrassed. It is almost like by being here I am admitting failure. I'm scared since I have never had an MRI. I hear it is like being a cigar stuffed into a cylinder. As I approach the reception desk I catch myself starting to intellectualize things.

"Hi, I'm Dr. Donaldson," I say majestically.

The receptionist looks at me and says ever so politely "Please fill out these forms. If you need help with anything I will be glad to help you." I don't think she intended it but talk about feeling belittled: after all I am the doctor and don't need any help.

I'm the last appointment of the day so I have the waiting room to myself. Peace and quiet with no distractions allows me to focus on my fears. I try reading a book and discover I'm too anxious to comprehend anything. Finally I am in the change room and then in the MRI. It is very reassuring to know there is a panic button especially as the machine sucks you into its vortex. The machine makes all sorts of sounds, none of which make any sense. It is nice to know you've been charged with enough magnetism to stick to the fridge on the way by.

I am finally done and get a chance to see what the radiologist sees. I see all these little round circles with black and white contrasts and lines and ridges and God knows what else. Making sense of any of this is beyond my skills. The technician sends me on my way with the comment that I should know the results in a day or two.

The single worst thing a person can suffer is waiting for the results. A day turns into two, turns into three, turns into four, and turns into a week. Maybe the radiologist saw something so pathological he needed a consult. Maybe he got hit by a bus? What is keeping him? Finally, a week later the results come. They are staggering.

The report reads like a horror story. First of all there are three fractures in the hip — two in the acetabulum, one in the head of the femur. In addition, there is a displaced fracture in the hip. I am stunned. How could all this be missed in the X-ray? No bloody wonder I am in so much pain. Anger is followed by a call to my doctor, a call to my wife and the spewing of venom to the Gods.

IF YOU'RE NOT PART OF THE CURE, YOU'RE PART OF THE PROBLEM.

~ ADAM KAHANE ~

CHAPTER FOUR

Pain - The First Six Months

~~~~~~~

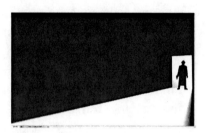

ONE OF THE MOST IMPORTANT YET HIGHLY AVOIDED sensations a person can feel is that of pain, for it is designed to protect us from the unpleasant experience of injury. People who don't feel pain often incur injuries that are life threatening. They don't know that there is something wrong, or that they are in danger.

When you put your hand on a hot stove, there is a signal that goes from your hand to your brain almost instantly. You remove your hand, consequently reducing the amount of tissue damage and pain. The pain signal is transmitted through nerve fibres to the dorsal horn of the vertebrae, up to the tactile and limbic parts of the brain. These nerve pathways are considered to be "hard wired" meaning there is little that can interrupt the signal.

When I fell, my hips and femurs (the bone at the top of the leg connecting hip to knee) sent a signal up to my brain saying, "We are injured. Do not move!"

Reflexively the muscles and other connective tissues around the joints went into spasm, through what is known as a "reflexive protective spasm." This mechanism, which is considered involuntary, meaning it happens without any conscious command, is designed to lock the bones in place in an effort to prevent further damage. A protective spasm wears off and also can be undone by will or force as seen with troops in battle, in sports, and in other situations.

Immediately upon falling stiffness dominated my hips, literally frozen into place by my muscles. I could wiggle my toes, ankles and knees but could not move my hips. After the fall, I was able to move my legs. Believing there were no fractures, Mary helped me up. Upon weight bearing there was a loud snap: the sound of the displaced bone moving back into its normal position. There was no pain with this so I assumed all was well.

This is what makes assessing tissue damage on the basis of reported pain so difficult, for even in acute pain factors other than tissue injury contribute to the pain report. Even though I had four fractures I felt no pain at that time, the brain fog masking everything.

Also it is known that not everybody reacts in the same manner to pain. Culture plays a big part in it, as does religion. For example the Scottish and English are known for keeping a stiff upper lip, while cultures in the Eastern Mediterranean regions are expected to verbalize their pain. The presence and acceptance of pain is woven into different religious beliefs such as "it is God's will." So pain reports are known to be unreliable as indicators of actual tissue damage.

Pain is a four-letter word that triggers more types of responses than any other word I know. It helps us survive, but most people remain anxious about it. It triggers emotional responses such as anger, anxiety, and depression. It triggers behaviours such as drinking, avoidance, rejection, medication/drug abuse and thoughts of suicide, hate and despair. Notice that most of these words are negative in connotation. There are none that indicate loving, caring or sharing. The most polite I have heard is "It is a cross I have to bear."

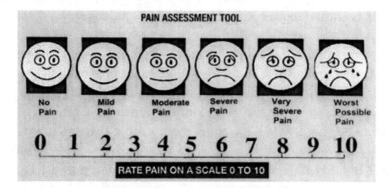

Despite knowing that the pain report is a poor indicator of problems, health care professionals still try to match it to the amount of anatomical damage to the amount of pain. The correlation between reported pain and amount of tissue damage has been researched extensively and shown to be approximately 20%. Thus we know from research that this model just does not work.

So what is pain? When you look pain up in a dictionary or online at Wikipedia, you get the following: "Pain, in the sense of physical pain, is a typical sensory experience that may be described as the unpleasant awareness of a noxious stimulus or bodily harm. Individuals experience pain by various daily hurts and aches, and sometimes through more serious injuries or illnesses. For scientific and clinical purposes, pain is defined by the International Association for the Study of Pain (IASP) as "an unpleasant sensory and emotional experience associated with actual or potential tissue damage, or described in terms of such damage."

Pain is also considered highly subjective. A definition that is widely used in nursing was first given as early as 1968 by Margo McCaffery: "Pain is whatever the experiencing person says it is, existing whenever he says it does."

Before the accident I had read the above passages a hundred times, knew and believed the IASP definition and applied it to my patients and their complaints. After all that was the universally-accepted definition, agreed to by people smarter than me, and one that could be used when I needed to protect my behind in writing reports.

Most health care providers consider pain to be acute for at least six months after an injury. At the moment of my accident I hurt in my hips and legs. I was angry, mad at myself for being such a klutz. My anger was not specifically connected to the pain. The pain was purely a sensory experience. Immediately after the examination by the orthopaedic specialist I flew to Miami to give a speech. I hurt and while the pain dissipated to some degree, it was still constant; it never completely went away. Weight bearing for any length of time was out of the question so I gave my speeches sitting in a wheel chair or on crutches. Life was becoming a process of adapting to the challenges of the pain; using crutches became the norm rather than the exception and ever so slowly I was restricting my activities and social contacts.

By definition, at the end of six months, my pain became chronic.

So, what happened to the pain between its onset and years later? It went from being acute to chronic. Unfortunately most people in the healthcare industry believe, and would have you believe, there is little or no difference between these two states (acute and chronic). This thinking pervades our health care systems, current populace concepts and medical-legal systems. Fortunately, current research is starting to break down these myths, but it takes at least ten years to get a new idea through the medical system.

WHEN ONE DOOR OF HAPPINESS CLOSES, ANOTHER OPENS;
BUT OFTEN WE LOOK SO LONG AT THE CLOSED DOOR THAT WE DO NOT SEE
THE ONE THAT HAS BEEN OPENED FOR US.

~ HELEN KELLER ~

CHAPTER FIVE

From Acute to Chronic Pain - The Chronology

~~~~~~~

THE YEARS FROM 2000 TO 2010 ARE A BLURRY, fuzzy recollection of events. Throughout this time I was dealing with gradually increased pain as well as psychological factors. I was also dealing with legal and health care issues and the various related professionals. This time period is not crystal clear in my mind. Unfortunately, I did not set out at the occasion of my accident to write this book so I did not make notes. After all, I expected to shortly be back on my feet. Most of the materials that follow are from memories and discussions. Within these limitations I am trying to figure out how I went from acute to chronic pain.

The progression from acute to chronic pain was an insidious process. There was not a moment I could identify as THE moment when I became a chronic pain patient. This change occurred over the years affecting several different aspects of my life.

Chronic pain just gradually wore me down, eating away at my soul, lifestyle, energy and ability to cope. I remember being a high energy person who at times became so tired that I did not think I had enough energy to lift myself off a toilet.

My patients relate much the same story. They report just gradually being worn down a little bit at a time especially if they are having issues with their doctor or an insurance company. As with most people, it is easier to identify problems external to themselves as opposed to looking within, so problems with others tend to be reported first before any others. Also, except for those that are suffering from a repetitive strain injury, most patients clearly can identify that moment in which their life changed – a life defining moment (i.e., motor vehicle accident).

Several forces work independently and collectively to create and maintain chronic pain. My experience was no exception. In this chapter, in order to make sense of these forces, I track the chronology of events as related to the chronic pain and related medical issues. In the next chapter I discuss other factors that influenced and shaped my pain.

Initially, the pain in early 2001 was fairly localized, affecting my right hip. Gradually over the years my right knee, then my left knee, then my left hip and finally my back started hurting. The pain varied as to type, intensity and frequency. As did all my patients, I expected to get better and return to my pre-accident way of life. This clearly did not happen much to my surprise and then my desperation.

The pain in my right hip gradually dissipated but never completely went away. The first thing that happens to most people is they stop performing any movement that is associated with the pain; I was no different. After my fall I basically stopped bearing weight on my legs, depending on crutches for ambulation. Over time my arms got stronger, my legs got weaker, but I never was really aware of this change. My expectation that the post accident pain would go away and I would return to my previous life style continued. This was not happening. The pain continued. For a while I was able to get off my crutches, walking short distances with a limp, but nonetheless walking. Over time the amount of unsupported walking decreased and in six years I was almost totally dependent on my crutches.

I commonly see this pattern of gradual decline in performance in my patients. When there is no improvement they get anxious and worried for there is always a suspicion that something has been missed or that they have something more serious. Like my patients, these thoughts raced through my head every time my pain got worse.

During the time immediately after the accident and for the next few years I received massage and myofascial therapy. I was fortunate to have some top-notch therapists on staff. Each appointment produced much immediate pain followed by several days of pain relief characteristic of the response of myofascial pain to these therapies. Thus I believed I was on the right track and recovery would soon follow.

I continued, however, to be in pain so I sought further medical investigations and comment from other health care professionals. After numerous consultations with my family doctor I was sent to various specialists each one offering a different solution. Each referral sparked a degree of hope and a degree of skepticism.

Much to my surprise, in 2004 my right knee started aching. With increased activity, the pain became an almost intolerable throb. At rest, fortunately, the pain decreased. I presumed that something had been missed in the initial investigations, but an MRI disproved this. These results suggested to me the pain must be myofascial in nature and I was probably compensating for the hip.

During this time period I decided to sue the owners of the establishment at which I had fallen. Of course being a large corporation they hired some of the best lawyers in Calgary. I have never been able to figure out how the corporations save money doing this, for they wind up paying the lawyers thousands of dollars to save minimal amounts. After several meetings with some very obnoxious lawyers I decided that all the jokes about them were true and I remember thinking how sorry I felt for their kids.

As part of the investigations I was sent for an independent medical exam to an orthopedic specialist. During the examination he pulled on my knee so hard something snapped. I experienced no pain for about 24 hours and then, wham! did it start hurting. Stabbing shooting pain right in the center of the knee, getting worse as I straightened my leg. Standing was virtually impossible; climbing stairs brought me to tears, lying in bed brought no relief. I resorted to taking Oxycontin. Myofascial release therapy kept me going for a couple of months, but the pain was unrelenting.

Finally I had another MRI done. This one was positive for a tear in the meniscus. I did not report this incident to the College of Physicians for I knew that getting a specialist to testify against a colleague was almost impossible. Besides I was starting to wear down and run out of energy.

Despite the pain I remained as active as I could in the community. I served as Director of Medical Services for the 2004 Summer Games. This latter event took over a year to plan and during that time I noticed a marked decrease in my ability to function. One event in June 2004 particularly stands out. The games were about to start and the staff was loading medical supplies into our temporary clinic. I discovered I could not even carry a flat of water. It was just too heavy and produced too much pain in the hip and knee. As I sat there and let others do the work, I realized how much I had deteriorated and that Mary was right, I needed to do something. Going back home I picked up a number of different objects only to realize I could not carry them. I could not go up and down stairs and could not walk a block. What a shock! Utter disbelief rocked the very core of my existence. I really was handicapped! I could not deny it any longer.

Earlier I mentioned how insidious chronic pain is. My present state had just slowly evolved without any dramatic events other than the one medical incident. It was just a slow, gradual deterioration in my functioning. I lost a little bit each day.

In 2005 my right knee was scoped, fixing the meniscus. The surgeon was kind enough to let me watch on a monitor. Meniscus has the same look and texture of scallops, thus affecting my appetite for this delicacy, although today my interest is slowly coming back. Immediately after the operation there was significant reduction in the pain in my right leg, increasing my hope for recovery. I hoped any damage to my knee had been missed in the initial investigation. Again, like the medical practitioners, I was looking for an anatomical explanation for the pain. But the hope only lasted for about six months until the pain came back.

As time went by, I was doing less and less in the house and in the yard, and started to withdraw socially from community activities. My energy levels went downhill rapidly. It took too much energy to cope with the pain, leaving nothing for anything else. I was an active member of Rotary International; club president in 2004 – 2005, but after completion of my term I almost completely quit. It just took too much energy to participate.

By 2007 I hit rock bottom, dependent on crutches to walk, unable to stand or sit for any length of time, depressed, withdrawn and feeling desperate. In September of that year, my world started to turn around. I saw an orthopedic surgeon who strongly advised me to have both hips replaced. While I respected his opinion, I needed some time to process this for I had so much pain throughout my legs I was not convinced the problem was in my hips.

During this time, the pain worsened when lifting and trying to carry a load of newspapers. I felt a horrible tearing of the ligaments on the inside of my left knee. Searing, agonizing pain gripped that whole area. Consequently, I decided to have my left knee scoped twice. The first time was in 2007, but did not get rid of all the pain. In consultation with the orthopedic surgeon, a second one was performed in 2008. As hoped for, a temporary reduction in pain occurred, unfortunately followed by a return to the pre-surgery pain level. In the fall of 2007 I decided to have the recommended hip replacement surgery. X-rays of my hips took my breath away; there was no capsule space in either hip and both had to be done. What was not explained is why this deterioration was not found in

2000. I had spent a better part of 55 years trying to put off hip replacement surgery, but the data was there in black and white. It was time.

Other than the chronic pain, my health had been quite good with little time off from work. I enjoyed the interactions and successes of my patients. Back in 2002 I had a complete physical and passed without problems. However, my pre-surgery work-up found I had low platelets and the surgery was cancelled. I needed this checked out as I could bleed to death during the surgery. I reacted with utter disbelief. How could I have a blood disorder on top of all the other issues I was dealing with? It seemed unfair.

I saw the hematologist just before Christmas. My platelet count was so low they decided to do a series of tests to rule out possible causes such as leukemia. I had to wait until January to get the tests done. This was the worst Christmas of my life. I didn't even want to put up the Christmas tree. I was in pain all the time, unable to function well and was facing the possibility of a terminal diagnosis. Time seemed to stand still. Seeing other people enjoy the holiday season made me want to sulk and drink. Fortunately, my strong-willed wife, Mary dragged me through. Sulking was not permitted. My dad's message, "Get off your ass and get going," kept running through my brain.

Finally January 7th, 2008 arrived. As I sat in the waiting room at a local hospital I had mixed feelings. There before me were horribly sick people, some fighting for their lives, all of them looked scared, frightened and bewildered. I was scared — shaking like a leaf inside. Outwardly I was quiet, imagining what each person was going through. Reading was out of the question. My mind was racing, going too fast. The staff was doing all they could with too few resources and too many sick people. I was also filled with a sense of humility; I didn't believe I was as sick as the others.

Finally, a nurse came out and got me, wheeling me into a private room with an exam table. They were going to do a bone marrow biopsy and I had no idea what to expect. First lidocane was injected into my skin to numb the pain. Positioned lying on my right hip, a large needle was stuck into the bone at the top of my left hip. The doctor wiggled the needle around causing a scraping sensation on that bone. Then the needle went through into the bone marrow itself. Other patients had told me this procedure was excruciatingly painful, but so far I had no problem. The needle was empty. It was used to draw bone marrow out of the hip.

The moment they started to withdraw the bone marrow I experienced horrific searing pain in my foot running up to my hip. I had never felt anything like it before and hope never to again. The nurse who was holding me kept telling me to breathe and keep breathing. A brown colored sheet flashed over my eyes as I headed towards passing out. I saw nothing but the sheet for the duration of the test, mere seconds, although it felt like forever. After the needle came out I lay on the bed soaked in sweat, disoriented and ready to die. After twenty minutes, as I left the waiting area, I did not feel sorry for any of the others any more. I just said a silent prayer for them hoping their pain would never be like what I had just gone through.

Every test they did came back negative much to my delight and the doctor's confusion. I was elated. I was going to live, but disabled with pain. To this day I am officially diagnosed as ITP (Idiopathic Thrombocytopenia) because all the tests remain normal. ITP is a condition in which the platelet count in the blood is low and the doctors don't know why.

I had an ugly encounter with an internal medicine specialist who diagnosed me with myleofibrosis. This is a terminal disease in which the bone marrow cannot produce blood cells. Fortunately a day after this consultation I saw my hematologist who called the diagnosis hogwash.

I do have a theory about the cause of the low platelets. I believe that in 2004 I was bitten by a hobo spider. I had all the marks including a puncture wound and a clearly identical pattern of skin redness as outlined in textbooks. The specialist did not believe that this could be a cause of ITP. Hobo spiders however, kill their prey by injecting it with venom that destroys the platelets causing its victim to quickly bleed to death. My general practitioner believed my theory and placed me on Tetracycline for one year as a precautionary measure. Today I still suffer from a low platelet count.

My left hip was replaced in 2008 and my right in 2009. I am now officially bionic. Modern medicine has done a great job in learning how to control acute pain. New medications, electrical techniques and surgical procedures all work towards reducing and in some cases, eliminating acute pain. After the hip replacements I was up on my feet within eight hours. By day four, there was virtually no hip pain. Unfortunately I continued to have pains in my legs, knees, and back.

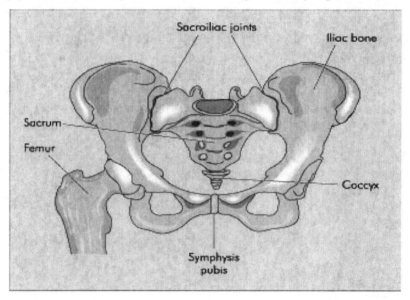

After surgery, I slowly and gradually improved as I relearned walking, moving, and doing the things I previously took for granted. My stamina slowly improved, my energy came back and I believe I became a better therapist.

I continue to have pain in my legs to this date, particularly the left one. For fifty-five years I had a significant leg length difference of 3/4 of an inch. The surgery corrected this, but in doing so has thrown off my motor control.

Toddlers learn and develop motor control. Motor control is the neurological process whereby the brain initiates and controls our movements. It allows us to move without conscious awareness.

I am now constantly scuffing my left foot because I have no idea where in space it is. My brain has adjusted to my body being out of balance with the right side lower than the left. To walk I learned to kick my left leg out to the side and then bring it around to the front. Now I no longer have to do this, but my brain is still back in infancy thinking I should. When I'm fatigued, the scuffing is particularly bad as my brain reverts to its old habits. My muscle movement pattern changed from what I had become used to back to the way muscles are supposed to work.

From this new walking style, I have pain in my left quadriceps that resembles that of a Charley horse with assorted pain in the related muscles.

This is heaven, though, compared to the way I was before the surgery. I can walk, albeit not for great distances, and I can sit, stand, sleep and enjoy life again.

Do what you can, with what you have, where you are.

~ Theodore Roosevelt ~

CHAPTER SIX

From Acute to Chronic Pain – Psychological Factors

~~~~~~~

AS I WAS EXPERIENCING THE INSIDUOUS ONSET OF CHRONIC PAIN I was also noticing changes in my psychological and emotional states. My training as a psychologist helped me be aware of many of these emotions and feelings. Some changes were very apparent as they occurred, while others only became apparent as I wrote this chapter. While these changes paralleled the development of the chronic pain and were part of the chronic pain experience they did not occur in any orderly manner.

I was/am suffering from what doctors euphemistically label "chronic benign pain." This means they acknowledged that I have pain — not caused by a recognized disease process, but by a process they couldn't identify. I was told that the pain wouldn't kill me, but what wasn't said was that it could destroy my life.

As I sat and contemplated this diagnosis I was not worried for I had studied and treated hundreds of chronic pain patients. I thought, "Just apply what you practice on others to yourself and the pain will go away." Then I hoped to return to pre-accident status. As time went by I started wondering why the pain was not getting better. I still continued to have success in decreasing my patients' pain, but my own pain level was not improving.

Doubts started creeping in; not doubts about my therapy skills, but doubts about my health status. Maybe something had been missed so I sought second opinions. Each second opinion came back negative, further confusing me and increasing my doubts. My strong unshakeable belief in the efficacy of biofeedback treatments for chronic pain and the ongoing research about chronic pain counteracted my doubts leaving me in a vortex of confused thinking.

As time went on this confusion became part of my life. My training in crisis intervention had helped me learn to live with ambiguities and confusion, so I put these confused thoughts in the back of my mind. I did however take extra precautions to make sure my patients continued to recover. My job kept me going, as I was able to lose myself in my work, maintaining my energy for the patients but gradually giving up attending conferences, preparing and giving speeches and not doing any more research. I just did not have the energy to do these extras after taking care of patients.

Over the years, without realizing it, I put up a psychological barrier between myself and the world. This helped me block out the pain but in doing so I blocked out everyone — especially Mary. I have a tendency towards depression, which I believe ran in my father's side of the family tree. My depression varied. While counteracted by memories of my dad's commands to get off my ass, there were long periods in which I would slip into feelings of hopelessness and reduced energy. These bouts of depression were clearly the result of increased pain and a decreased ability to perform tasks. When more energy was available it went primarily for work and for patients, with very little energy left for personal and community involvement. My pain went from being a simple sore hip to involving other parts of my body, to involving my emotions and activities, affecting all parts of my life.

The one thing I was able to avoid was drug addiction as my commitment to my patients was stronger than my need to be drugged by medications. I did drink alcohol but was never addicted to it. Throughout all of this my medical doctors remained supportive and helpful giving me medications as requested. I could go in and discuss with them what was happening never feeling that they doubted me, or what I was experiencing. I cannot tell you how important this support was.

Even more important was the support from my wife and friends. In psychology class I was taught how important a social network is in combating depression and suicide. This experience taught me just how true this is. They never doubted my reports of pain or my emotional outbursts, and continued to give me unconditional support. When the pain got too much I would get grouchy, which was quite apparent to anyone around me. Fortunately my sleep remained good, with only the occasional need for medication.

I occasionally thought about suicide but it was never a real consideration. Given my experiences in crisis work, despite all my problems and pain, I believed in a time of an improved life. I knew that biochemically depression and chronic pain share the same co-modulators meaning that as one changed the other would change, so I maintained my focus on reducing my pain not worrying about the depression.

My inability to do simple tasks around the house put a horrible strain on my family, which created emotional turmoil. When in pain I could not do the things I used to do. It forced Mary to do things she was not used to doing. As my role changed I became anxious, grumpsy, short tempered and was building a wall around myself so people could not see how much I was hurting and how frustrated I was becoming.

Chronic pain is a stimulus that occurs all the time. Imagine taking your busy schedule and adding to it something that demands your attention 24 hours per day 7 days per week. Every time my pain occurred, it required my focused attention to deal with it. This left me exhausted. Some nights it was all I could do to go to bed. As a human being I only had so much energy, and on numerous occasions I just plain ran out of gas.

On top of all of this is the loss of time that the doctors' appointments took up, and treatment sessions devoured. My life became a circus of appointments and therapy that affected my home and professional lives. Attending appointment after appointment and sometimes waiting for hours added to my exhaustion. I became increasingly pissed off and angry as I got caught up in a cycle of hope and disappointment, only to be dashed until the next appointment. I had to reduce the number of hours I could work, reduce the number of patients I could see and all of this created serious financial strain on my company.

I think back now on those years and except for the issues related to the pain, I have very little recall of events, parties, day-to-day occurrences, and anything else. Pain demanded my attention, forcing its way into my consciousness and forcing out all other awareness. This realization only occurred as I was writing this book and quite frankly scared the crap out of me.

1   Three processes, which evolved over time, were significant in helping me cope. First I made a game of forecasting the weather. Chronic pain patients regularly tell me how their aches and pains would change just before a storm front arrived. This was particularly true for headache sufferers, as their headaches would dramatically get worse. So forecasting the weather due to changes in my aches and pains became my hobby. Cold weather remains a torture for me as every part of my body, especially my legs, stiffen up.

2   The second thing I did was continued to read about new ideas and treatment techniques for chronic pain. Each technique gave me the same hope and bout of new energy of my patients. I was and still am introduced to new ideas and therapies that claim to reduce pain. Chronic pain wins when hope is lost. Throughout all the adventures with chronic pain treatment I have never lost hope, but in hindsight shake my head at some of the things I tried.

3   The third factor that emerged was my increasing awareness of the role that the brain plays in chronic pain. Several research papers suggest that chronic pain actively involves the brain, actually changing the brain's patterns of activity. This was exciting for me, as we had been treating the brain in our chronic pain patients for years. For me personally I was so focused on my body I virtually ignored what was happening in my brain and never did treat it. Without my awareness, my brain had to remain active, alert and involved in the care and treatment of chronic pain. Also, I believe that every patient has his/her own unique story to tell. Recognizing the subtlety and differences in their stories requires active listening, focused concentration and mental gymnastics, which kept my brain focused on something other than my own pain. In essence my patients were treating my brain. As important as hope was to my survival, the role of my brain became equally important.

Today I am in much less pain. The remaining pain is in differing locations in my body depending upon what I am doing or have recently done. I believe most of it is muscle-based. Even as a specialist I have not figured out the cause or process involved in chronic pain, whether others or mine.

When pain lessens it indicates that our intervention is having a positive effect and the treatment is accurate as to cause. As pain resolves it goes from being generalized throughout a region to becoming more focused in specific locations. I used this as a personal marker for improvement, as well as noted the changes from constant to variable and the number of days in which I won and the pain lost increased.

I am however, overly sensitive to the feelings occurring in my body as I am always worried about the pain returning. And there are sounds and sensations occurring regularly. Trying to figure out what is normal versus what can cause me pain creates numerous interesting dialogues between the part of me that worries and the part that tells me this is normal. This is a process that I share with all my patients.

Today I believe I am more emotionally stable than before. I'm no longer depressed: my doubts are gone and my energy levels have returned as evidenced by the writing of this book and my return to the lecture circuit. I believe I am more humble as a therapist understanding in greater depth what the patient is telling me. Most of all I hope I am more sensitive to the needs of others.

GOD GRANT ME THE SERENITY TO ACCEPT THE PEOPLE I CANNOT CHANGE,

THE COURAGE TO CHANGE THE ONE I CAN,

AND THE WISDOM TO KNOW IT'S ME.

~ AUTHOR UNKNOWN, VARIATION OF AN EXCERPT FROM

"THE SERENITY PRAYER" BY REINHOLD NEIBUHR ~

CHAPTER SEVEN

Medicine

~~~~~~~

THE SWEAT FORMS A RIVER as it collects between my shoulder blades and runs down my back. I struggle to maintain my balance. It is a clear bright sunny day with a temperature of minus 14 degrees Celsius as I cross the ice field. On crutches I pick my way through slippery patches of ice interlaced with ruts of packed snow and frozen mounds of snow. I inch my way along in trepidation; I'm terrified that I will slip again. Every step is one of focused concentration as my crutches want to skid out sideways and I do not want to put any weight on my right leg. I am terrified of falling for I can just imagine landing on the four fractures in the right hip, making them even worse. My imagination has me lying there with no one noticing me slowly and painfully freezing to death in a sea of pain and sweat.

Finally after an eternity (about 20 minutes) I made it to an area that is free from snow cover. I look back at the path I had just taken with a feeling of accomplishment like I had just conquered Mount Everest. I had crossed the parking lot, now on to the doctor's office.

I pass through the main doors of the medical center to be greeted by a large reception desk. Behind it sits a woman busy with paper work. I announce to her that I am Dr. Donaldson and I have a 1:00 PM appointment with Dr. X. Without looking up she hands me numerous sheets of paper to be filled in and taken with me to the examination room, room 201 down the hall. I swear if she had smiled she would have fractured her face.

Down the hall I go, trying to walk on the crutches while carrying the numerous sheets and trying not to drop them. Trying to wrap my hand around the crutch while not wrinkling the papers was a challenge, yet I plodded down the hall. I had expected that room 201 would be close to reception. As I passed a room labeled #10 and the next one labeled #11 my hopes started fading. It seemed that the hall went on forever. I trudged and made it in about another 20 minutes, having to pick up papers only twice.

Room 201 was a moderately sized room painted in faded yellow, with an examination bed, a chair and an X-ray screen. After a brief wait the doctor came in, sat down, explained that he was a resident and asked if it was okay if he did the examination and present the findings to the other residents. I appreciated him asking me and of course, agreed to everything he said.

The examination went smoothly with the resident trying to turn me into a pretzel, testing the various aspects of my range of motion for both hips. I never realized I could be put into such unusual positions and not feel any pain. Of course weight bearing was out of the question but he had me do it anyway. Finally upon completion of the examination the head orthopedic surgeon and about ten residents walked in to discuss me. After about a 20 minute discussion in a language with which I am not familiar the head surgeon congratulates the resident and turns to talk to me. He stated that it had been too long since the accident. I should have had surgery within a day or two of the accident but now the risks involved outweighed the rewards. Give the wounds time to heal and take care of myself.

I left Room 201 elated, as I did not have to have surgery. The trip down the hall and the frozen face at reception did not bother me at all. I stepped back out into the bright, sunny, chilly day, everything heightening my senses and compounding my sense of relief. The exhilaration lasted but a moment as I lay my eyes onto the ice field between my vehicle and me! I decided then and there whoever designed health care facilities; especially access to orthopedic centers must be sadistic in nature.

The most important person in the health care chain is the physician. He can either help or be a hindrance to your care. Nobody else's word carries as much weight to the patient, to insurance companies and lawyers, so the potential to do much good or much harm rests here.

Hopefully, most chronic pain patients have a family practitioner. Just after the onset of the pain, contact is frequent, relying heavily on medications to reduce the suffering. When medications don't work or stop working, other resources such as physiotherapy may be tried. Support by the family doctor is usually quite variable, with some doctors remaining supportive, while other doctors want practitioners in other disciplines to take this patient off their hands. As a result the treatment of chronic pain patients may often be disorganized and inconsistent.

As time goes on the chronic pain patient is usually sent to a specialist, maybe two or three. When nothing comes of this she often finds herself on the receiving end of further prescriptions and left to her own devices often being told it's in their head. Many search for alternative treatments.

I went through some of this, seeing specialists and the like, but as a practitioner, I knew what to expect from the system. By my design, I had very supportive family practitioners. I only sought out those who had philosophies similar to mine when it came to life and pain.

My first contact with any physician was the waiting room. Waiting rooms are very interesting as there can be virtually no one there, or there may be the proverbial mass. Personally I hate crowds, so needless to say I hate crowded waiting rooms. In them I would distance myself from others by retreating into my head. There I would guess what disease or condition brought people in and hoped it was not contagious.

When entering a waiting room, I egotistically expected the professional courtesy of getting in right away. As Dr. Donaldson, some offices recognized me and got me in right away, while in others it was hurry up and wait. (For an entertaining example of expecting professional courtesy and not getting it see the movie *The Patient*.)

An interesting dialogue usually developed when I had to wait. Part of me would defend the practitioner recognizing how emergencies happen and how needy patients would back things up. When I was tired, hurting, pissed off, and at times a little scared I would stew when kept waiting. When waiting for a specialist, I would think maybe they will discover the missed something that could have been cancer and is probably terminal, etc. I would never share these thoughts with my wife or staff for I felt they had enough to worry about. I didn't think they needed to add my fears to their worries.

The ambiance of the medical office also affects the patient's presentation. A warm nicely-decorated office usually left me feeling secure, while a sterile cold office made me feel insecure. Because of this I make sure my own office is nice and friendly. Research shows that a nicely decorated or warm office elicits a realistic pain report; the colder the office the more the pain report is exaggerated.

The personality of the physician dramatically alters the way the chronic pain problem is presented to them. In my experience, general practitioners tend to be more personable than specialists. Research also suggests that as the patient goes up the medical hierarchy, the greater the exaggeration of the symptoms occurs. For example the presentation of the symptoms in a family practitioner's office is less severe than what is presented in the neurologist's office. Knowing this I kept consistent in the reporting of my pain.

A warm comforting physician will elicit more emotional complaints than one who is cold and aloof. Thus the supportive family practitioner gets overrun with chronic pain patients. One of the great mysteries of life is how do chronic pain patients find out which doctor is supportive and which isn't, for chronic pain patients will flock to certain doctors.

Notice paperwork was left until the bottom of the pile. I hate paper work. If I filled out one set of paperwork forms I must have filled out 20 sets, every time getting a little more pissed off. While I understood the need for the information, I also considered it an invasion of my privacy.

I do not know why everybody's forms need to be different, but I have the fantasy that one form should suffice when asking for name, address, medical history etc. Just photocopy that part of the paper work, hand it to the receptionist and then complete the section that is unique for that particular specialist.

THE MEDICAL MODEL

As a specialist in chronic pain I would often think that I knew more about chronic pain than the clinician I was seeing. Only a few people impressed me with their knowledge and their willingness to spend some time with me to share it. Unfortunately at our clinic, we only see the treatment failures giving me a jaundiced view of the health care system. This left me a skeptic, as I knew the medical model limits physicians' understanding of chronic pain and their ability to treat it. On the other hand I was in pain and really did not care about these issues. Consequently I would wind up trying to get all the help I could, while remaining a skeptical, hesitating, questioning, pain-in-the-ass patient.

The medical model assumes there is a direct and high correlation between the amount of tissue damage and the reported pain. This clearly does not hold up for acute pain and is even worse for chronic pain.

The studies with which I am familiar indicate that the correlation between tissue damage and reported acute pain is about 20%. This means that only in 2 of 10 cases the amount of tissue damage matches the reported pain. The relationship is so low due to several factors. Culture, social context, personality and self-awareness all play a large part, as does limited knowledge of the human body.

Each culture has its own recognized way of expressing pain. The English and Scottish (my heritage) are famous for keeping a stiff upper lip. Conversely people from the Eastern Mediterranean are taught to express the pain, often quite graphically. No method of expression is correct; they are what they are, a culturally defined and accepted method of verbalizing (or not) pain.

My cultural background and training taught me to minimize the pain, masking its seriousness. My stiff upper lip after the fall and in the emergency room after physiotherapy are examples of how I may have confused the attending physician and led him to an incorrect diagnosis.

Secondly, The pain report itself, while being essential to direct the health care provider, may be so confusing that it, in and of itself, may lead to missed diagnoses.

Lastly, the social context or situation in which the pain occurs is important. We have all seen the sports hero who plays through the injury and is adored by the team and sportscasters. The same injury in another situation would probably require an ambulance. One example of this is a concussion sustained during sports activities that show up years later as severe cognitive difficulties.

The context in which the chronic pain starts is important in determining what happens immediately in treatment but may be irrelevant later on. The most common example seen at our clinic is a normal healthy individual that is injured in a car accident. In one moment an entire life is turned around, going from being routinely normal to highly traumatic. Years later the cause of the trauma is not important but the after effects are now the issue. This time element is a critical piece in understanding chronic pain and in dealing with medical professionals and insurance companies.

An individual's personality plays a part. If a person is or has been rewarded for being sickly, by others, this will impact the pain presentation. While denying it, there were times in which it was easier to claim pain than force myself to do something unpleasant.

A history of physical abuse also appears to play a part in chronic pain. The incidence of people who have chronic pain, and who have also been physically abused is quite high. Some of the studies that I am familiar with indicate it is in the 40% range. While the exact mechanism is unknown, it is believed the neural pathways involved in pain are sensitized to the its transmission, leading to the development of the chronic pain. Basically old dormant pathways, , are re-activated by new pain signals, thus requiring less pain stimulation to be involved in the chronic pain cycle.

Finally self-awareness plays a major part in the chronic pain puzzle. By self-awareness I mean knowledge and awareness of our body and its associated sensations. At any given moment, say right now as you read this book, your body is alive and performing different functions or tasks without you having to direct it to do so. You might be feeling your heart beating, gas pains in your bowels, your mind racing or pains in your muscles. For example, pain in your buttocks muscles alerts you to the fact that you sat too long in one position and it is time to move.

These issues get muddied in the chronic pain situation; what if the signal that you are sensing is actually pain due to your injury acting up, or even worse, what if you have made the injury worse? Individuals who are highly aware of their body suffer greatly because it is difficult to differentiate between those body sensations that are perfectly routinely normal and the chronic pain. To be on the safe side most chronic pain patients assume it is due to the injury. A large part of my job is to help people separate normal body sensations from those sensations related to the pain. Fortunately for me, I learned kinesthetic awareness during my years of crisis work and did not experience this problem to any great extent.

While tissue damage presents differently, it's report can be the same as for muscular pain. This is why medical practitioners rely heavily on objective evidence (i.e., X-ray) to determine the problem. You can imagine the problem a physician has when pain is reported there are no objective markers.

A major problem with the medical model in this treatment is the lack of emphasis they place on muscles and myofascial problems. This is created by two factors: a) the lack of tools to objectively diagnose muscle dysfunction and b) most doctors are not trained to recognize trigger points and tender points.

A start in this battle to ascertain objective measure is the development of surface electromyographic (SEMG) techniques. While they are routinely utilized in the pain clinic, much research and standardization of these techniques is needed.

Immediately after the accident an SEMG assessment showed that my right side gluteus maximus (buttocks) and iliopsoas (inner hip muscles) were not working well. Over time, this was corrected with SEMG therapy but these muscles also needed strengthening. Getting myself motivated to exercise is a struggle at the best of times, and it certainly didn't help that the pain made it difficult to do so.

Due to the pioneering works by Janet Travell and David Simons, as well as the work by many rheumatologists on tender points, knowledge concerning chronic muscle problems is slowly changing. However the dissemination of information in all health care fields is slow, taking years for ideas to be accepted and respected. Forty percent of the human body is made up of soft tissue, yet problems with it are often missed, misdiagnosed or even worse, assumed to heal in six weeks. This latter assumption has led to immeasurable suffering for those who have sustained a soft tissue injury.

Mainstream physicians have basically two tools with which to treat: drugs and surgery. As I mentioned previously I was up and walking eight hours after my hip replacements. Wonderful strides have been made in the alleviation of acute pain. Unfortunately this cannot be said for chronic pain. What is known as the *Gate Control Theory of Pain* focused drug research and the treatment of acute pain into the areas of the spine known as the dorsal horn, a grey matter section of the spinal cord that receives several types of sensory information from the body.

The Gate Control Theory suggests that the spinal cord contains a neurological "gate" that either blocks pain signals or allows them to continue on to the brain. It differentiates between the types of fibers carrying pain signals. Small signals go through where those sent by large nerve fibers are blocked. This is often used to explain chronic pain.

Thus numerous acute pain medications were developed which blocked the pain signal at this site providing major relief from acute pain. Also major advances were made in the use of narcotics for the alleviation of acute pain. The assumption was made that if these drugs work so well for acute pain they must also work well for chronic pain. WRONG. These drugs do not work well for alleviating chronic pain and represent a major problem for medical science They have not as yet been able to develop a method of pain control that works on chronic pain, with current techniques providing only temporary relief. The fact that the drugs for acute pain do not work for chronic pain suggests that there may be other neurological mechanisms involved in causing and attending to chronic pain.

As mentioned one of the major tools physicians have in treating physical problems is surgery or forms thereof. However, in general, surgery is not used in the treatment of chronic pain. Nerve blocks and acupuncture work in select cases, but the research literature on these treatments is still evolving.

You Could Not Pay Me Enough To Be A Physician

It is not my intention to belittle the medical profession. I have said many times you could not pay me enough to be a family practitioner. Think about it from their point of view. We ask them to diagnose a problem that with present science cannot objectively verify in five minutes or less (the average time most physicians spend with their patients), and then they have to resolve the problem, often resulting in pain medication prescription without making us addicted. Also, on top of that, they have to make sure we are not otherwise seriously ill or worse and not just doctor shopping for drugs.

Remember physicians are human. They want their patients to get better and when the patient does not get better, they often view it as a failure. I have nothing but upmost respect for most physicians, especially the family practitioner.

My experience in seeing the specialist is quite different than that of seeing the family practitioner. The specialist's offices tend to be located in the farthest corners of a building with limited accessibility.

I struggled on my crutches to get to the building for one of my visits. I had sweat running down my back again. This time it is a clear bright sunny day with a temperature of plus 15 degrees Celsius. The nearest handicapped parking left me parked a block away. I arrived at the office hot, smelly and tired, to be greeted by a pleasant young woman who asked me to take a number. Looking around I see a room full of what I perceived to be injured, angry, tired and fed up people. It took me a minute to realize these were patients, not the staff. Then I realized this line was to see the receptionist, not the line to see a doctor. I was number 122; at that moment they called #68 to the reception desk please. Two hours of sitting waiting for your number to be called passes very slowly. Yet, I felt my pain to be insignificant considering those around me.

Finally it was my turn to get registered with the nurse who said rather unconvincingly, "The doctor will be with you shortly." I am placed in an examination room and proceeded to wait. 'Shortly' turns out to be 30 minutes. And with each passing minute I got angrier, fed up and wanted to leave; I thought amputation of the hip would probably be quicker and easier. The door opens and I immediately adopt my stiff upper lip routine, minimizing my complaints and pain. The specialist takes a quick look at the referral letter, pretends to listen to me and says, "You need an X-ray. See the nurse for the requisition." He leaves: total time two to three minutes. I am stunned, I am angry, I am frustrated and now I am acutely homicidal.

Why did I waste all this time for a two-minute meatpacking assembly line consultation? I decided I really didn't need this second opinion and left without picking up the requisition.

Specialists focus on specific physiological systems: that is how they are specialists. However I have thought several times that they forgot that I was a human being. This behavior obviously didn't satisfy me. I wanted to know what the specialist was thinking, and I needed some emotional support. I just want to remind specialists: remember there is a real live person sitting across from you, who is probably scared and/or experiencing some sort of emotional reaction. In the end I found an orthopedic surgeon who communicated well with me, reducing the emotional impact of hearing that surgery was needed.

There is one group of health care workers that I have absolutely no respect for. These are the people that tell you: *it's all in your head*. I take great exception to this labeling for several reasons. First of all, it is a diagnosis by exclusion. This means the doctor is making the diagnosis on the basis that they can't find anything physically wrong. So instead of saying that, they say it's 'psychological.' They would not diagnose a broken leg without doing an X-ray, so why do they come to this conclusion without doing any psychological testing? I believe they make this statement because their ego won't allow them to say they can't find anything wrong. Why can't they just say "Within my area of expertise I can't find anything physically wrong to match your pain report?"

Also, physicians took an oath that starts out "First do no harm ..." By telling the patient that what they feel is psychological they cause harm. This often produces severe emotional upset leaving the patient questioning his or her own senses and feelings. Much time is spent in my practice dealing with the damage that this statement has caused. The point is, it may be psychological but please leave it to the psychologists to sort that out.

I say a sad "Amen!" to the above. Remember the trip I had to the emergency department after receiving interferential stimulation from the physical therapist? Any physical therapist that uses electrical modalities will tell you that applying it across a broken bone will cause severe pain. The trip to emergency occurred between the X-ray results, which were negative, and the MRI, which showed four fractures. The response of the staff at the emergency department was belittling and insulting. Nobody should be exposed to that attitude. I remain pissed off about it, even though I still work with some of them. It was hard enough to deal with the pain, not knowing what was going on, but then to be told it's in your head, take a pill, relax and without saying it in so many words quit bothering us – we need your bed for someone who is really sick.

Dr. D's Advice for Working with Your Physician

Remember your physician is one of the most important people in helping you cope with and hopefully eliminate your pain. As such I recommend you do the following:

1. Keep your message to your doctor simple and straightforward. When I went to see a doctor I would mentally rehearse what I wanted to say. If you can't remember things write them down. Your message needs to be clear, succinct and to the point. If I wanted extra time to discuss something I would let the receptionist know so she could adjust the schedule and I would not feel like I was interrupting a busy schedule and overflowing the waiting room. Doctors are in the medical field because they want to help people, but they also run a business.

2. Doctors try to base the diagnosis on objective data (i.e., X-ray) trying to match the pain report to anatomical knowledge. The pain report advises them where to look for anatomical problems. If you tell your doctor you hurt everywhere, where are they going to start looking?

3. Tell your doctor if you have noticed anything unusual or any recent changes. It helps your doctor if things can be tracked over time producing a pattern.

4. Tell your doctor where you hurt, when you hurt and what you cannot do i.e., bend over. Do not go into a long description of what is wrong with your life because they will not remember all of it. Like the rest of the human race they usually only remember about the first 20 seconds of a conversation.

5. Answer your doctor's questions as directly and succinctly as possible. Again no lengthy speech on how badly you hurt.

6. Remember if you keep talking about how bad the pain is you will get a prescription for a pain killer because your doctor does not want you to suffer. Often I would refuse these. In doing so I fantasized that the doctor felt like he/she was not doing their job.

7. Remember your doctor is likely trained to use medications to solve problems. If your doctor keeps giving you pills and more pills with poor results make sure that he/she knows the medication is not working. Your doctor may want to try another type of medication. You are well within your rights to ask him/her if you can be referred to someone who works in a rehabilitation field like physiotherapy, chiropractic or psychology. You need to take the initiative to get a referral to an appropriate agency. Don't wait for your doctor to think of it.

8. You, as the patient, have to take control and make sure your doctor knows what is going on. When a prescription he has given you is not working, you need to let him or her know. If you feel you're just given a prescription and whisked out the door, you need to either discuss your feelings with the doctor, or find a new one.

9. If someone asked you how many times you have seen your doctor, how many prescriptions you have been given and whether you've reported the results of taking the prescription back to the doctor, would you know? Most people can answer only parts of these questions, and it's up to you, as the patient, to know these things. Both you and doctor need to know exactly how your treatment is working, and what can be changed. Caution: Don't get caught up on keeping records and tracking every last thing surrounding your condition. This kind of compulsive record-keeping only serves to keep you focused on your pain — and not on getting better.

10. Open and honest communication with your doctor is vitally important. If the doctor's reaction to your report is negative, then you need to be looking for a new doctor, one that is willing to see chronic pain patients. Unfortunately today there is a shortage of doctors willing to do so, so finding a new one may be more difficult than putting up with the old one.

11. Finally a couple of words about the Internet. The Internet allows you to be knowledgeable about issues pertaining to your chronic pain. It is both a valuable source of information and a curse at the same time. The value lay in that a wide variety of information is now available to all people. By doing careful research, in all probability, you will know more about factors involved with your specific chronic pain than your doctor does. Use this information to assist your doctor by assembling written materials and then discussing them with him or her. Some doctors will value this while others will be threatened. Do not tie up your doctor's time; pre-book any discussion time with the receptionist. *Mary and I spent hours on the Internet looking up possible solutions for my pain, each solution offering hope. And hope is so critical to the sufferer. When the pain was most damming I had enough problems just seeing all my patients and doing the necessary paperwork, let alone reading copious amounts of material from the Internet. I often thought the only time I could do this reading was between midnight and 6AM on Sundays when I had some spare time. Your medical doctor is the same. There is just too much information on the Internet for the health care professional to read about every condition they may see in a day.* You can take responsibility for the information you give to your doctor and hopefully it will get read. Of all the articles my patients bring me, I can read in detail about one article per week, and the others just get skimmed.

12. The Internet can also be a curse. Incorrect or inaccurate information is dispensed without

consequences. Individuals can use the Internet for profit and gain, using the hope of the sufferer to prosper. Also, especially on chat lines, severely negative comments and problems are posted. After reading one chat line about problems occurring post hip replacement surgery, I stopped. It was just too negative and upsetting. Most people who have had successful outcomes don't spend time in such chat rooms. They are too busy getting on with their lives and not focusing on their pain.

13. Also remember the mainstream medical professionals are trained to understand anatomy, disease pathology, and the use of drugs in treating various conditions. Do not expect them to know all there is to know about the various alternative therapies.

14. Your doctor will order various tests and images. Waiting for test results is the most anxiety-provoking situation that the chronic pain patient goes through. The health care system is well aware of this and works to process results as fast as possible. However, unless there is an emergency request the results are batched and processed, so individuals are kept waiting until an entire batch is done. Remember the lab makes money per test, not by sitting around doing nothing.

15. *As you may remember, at one point, right before Christmas, it was thought I might have leukemia and the tests were scheduled for January. Time has never passed so slowly as in that time period between Christmas and January. The answer was to do anything possible to take my mind off of things, to remain active, to go to the various Christmas parties, to take the dog for a walk, and play scrabble for hours at a time even if Mary usually won. This time is also equally stressful for the spouse. Mary and I had been married over 30 years. She is my best friend, so we spent a lot of time doing things together, which helped us both. We learned more about leukemia through the Internet than I care to remember, for us one way of coping is to know what we are dealing with. The results came back negative in all respects. I am sure you can appreciate our relief. In hindsight the thing that kept us going was keeping busy.* I can't emphasize enough the need to stay busy. Do not sit around and watch TV and don't feel sorry for yourself. It is not going to help you.

Advice For the Practitioner

1. Every once in a while I will have a patient referred to me who has extensive detailed notes about their doctor, their outcomes and their failures. One day a person came into the office with a binder full of 25 pages of notes pertaining to the care with various health care providers. All of their failures were documented. I now won't take patients like this. Early in my career I set out to prove how much better I was than any other professional they had seen. I too became a failure and I don't want to become another in their book. This kind of record keeping may reflect an obsessive-compulsive disorder and should be managed by a psychologist. Additionally, the focus is on the chronic pain and not on getting well.

Current research recommends that the treatment of chronic pain be multi-modal in nature as several different neurological systems are involved. Thus it is hard to believe that one pill will work for a multiple system problem, with medical science now looking at the administration of two or three drugs simultaneously. Also these same experts are recommending programs that involve more than just medications. The recommendations include things like acupuncture, physiotherapy, chiropractic, biofeedback, neurotherapy, exercise therapy and the like.

LEARN FROM YESTERDAY, LIVE FOR TODAY, HOPE FOR TOMORROW.

~ ALBERT EINSTEIN ~

CHAPTER EIGHT

Drugs

~~~~~~~

THE LIGHT STREAMS THROUGH THE WINDOWS on a cold winter's day; the fireplace sends its radiant warmth across the room replacing the chill in my body with bone/soul nourishing heat. The world is alive with sights and sounds: the howl of the whistle of a train in the distance, the crunch of snow that is so cold it has crystallized, the sight of your breath as it solidifies in the cold air, and the sound of coyotes answering the train. Beautiful. All of it was so beautiful.

The pain is resting in my hips ready to attack my serenity at any given moment or with any movement. But I don't care, for I am stoned, quite stoned. On Oxycontin. The pain didn't disappear, it still remained in my body and as long as I didn't move the euphoria provided by the medication helped me cope. What a relief from that nagging, gut wrenching pain. I drift off into a peaceful sleep.

Oh, am I hung-over. No I hadn't been drinking any alcohol; I had just taken one Oxycontin ten hours earlier. I try to fight my way out of a brain fog that leaves me reeling, unstable on my feet, slightly nauseated and needing to clear my head. I am basically unable to function, although I am awake and can speak but not too intelligently.

Drugs represent a two edged sword for the chronic pain patient: pain relief versus addiction or side effects. At times I wanted and needed the pain to go away for I needed a break. I was aware of possible side effects but really didn't care. Conversely this nagging voice in my head kept me very worried about addiction or worse.

Chronic pain can control your life, limiting movement, enjoyment, and your relationships with others; it can suck the life out of you. Painkillers give you relief from this. I was no different than anyone else wanting to escape even if it was only temporarily.

As a practicing psychologist I am aware of medications, their use, limitations and side effects. It is interesting today to watch TV commercials promoting the benefits of the use of certain medications. The commercial usually shows someone happy, active and enjoying life. Meantime there is this quiet voice overlay that usually goes on for 15 to 30 seconds warning people about possible side effects, including death. These commercials clearly portray the dilemma facing the chronic pain patient. On the one hand you need the pain relief. I wanted to be that happy carefree person in the commercial, just to have a few minutes of not hurting. At the same time I worried about all the side effects, addiction, death, and anything else that would add another problem to an already complex situation. So with this knowledge and fear I would go to the family doctor and discuss the issue of medications.

For people that don't know about the medications they are taking, learn them. Go on to the Internet and read, read and read. Print what seems important and take this to the doctor's office with you. Make sure that your doctor knows about all of the medications you are taking. Being well educated will help the doctor as well as reassure you.

Throughout the years of trying to cope with and lessen the impact of the pain I was exposed to a number of different types of medications. There are numerous types of medications on the market designed to control acute pain. As mentioned these generally do not work well with chronic pain. These medications include a) pain killers (narcotic and non-narcotic), b) anti-depressants, c) anti-inflammatories and d) others.

Only once was I incapacitated enough to take a narcotic pain killer and I was sick for hours after ingestion. Regularly I would use Tylenol because my blood disorder prevented me from using anything else. The blood disorder also precluded me from using any anti-inflammatories because of their blood-thinning actions. I personally never used anti-depressants but did see both good and bad effects in my patients. I refused to try any drugs that I thought did not have adequate discussion in the literature. Despite what the drug companies would have you believe the outcome(s) studies on anti-depressants are equivocal meaning that there is no clearly demonstrated advantage of one drug over the others. This leaves the doctor in a position of making an uncalculated guess as to what will work for you.

Previous experience had taught me that Valium worked wonders for getting rid of my pain and calming my nervous system down. However Valium is highly addictive. I worried about this because the last thing I wanted was to be addicted on top of being in pain, but some days I didn't care because I had so much pain.

In conjunction with my doctors I felt I knew enough to keep the Valium addiction from happening. Please note that I am not an expert in addictions and am not pretending to be. There are several factors that go into forming an addiction that I was aware of and used to guide me through this problem. First there appear to be genetic factors involved in forming addictions. In my case the family tree did not show any members with addiction issues. This family tree showed a tendency towards alcoholism particularly involving Canadian (Rye) whiskey.

I knew what is known as the "half life "of Valium. A half-life is how long it takes for a body to excrete or burn off half of the dosage of the medication. For Valium the half-life is four hours, meaning half of the dosage is out of the body in four hours. Taking the Valium at nine p.m. means that half of it was gone by 1:00 a.m. and 75% out by 5:00 a.m. The effect of Valium is to help me relax, calm my nervous system and help me sleep pain-free. This was achieved through judicious use of this medication.

With certain medications, one can get what is called a rebound effect. When you first take a medication it works well to decrease the pain. However as the effect wears off, the pain returns often to a higher level than it was to begin with. The reason for this is not clear, but rebound effects are well-noted in the research and addictions literature. I avoided rebound effects by sleeping through them when they would likely have occurred - at 1:00 and 5:00 a.m. Also if I became concerned about any possible risks or side effects, I would not take any on the weekend. Occasionally rebound effects would wake me up and more medication was not an option. Then it became a matter of suck it up and tough it out until the pain settled down.

The fourth issue is that of re-enforcing the pain. What happens is people take the medication to get rid of the pain. Thus the pain acts as the trigger for the medication to be ingested. So when people get the pain they take the medication starting the cycle of pain and medication. The medication wears off leading to more pain and therefore more medication. The way to avoid this problem is to take the medication only at a prescribed time, regardless of the level of pain. During the week I only took the Valium at 9:00 p.m., no ifs, ands or buts about it.

Valium generally takes about six weeks of constant use before a habit or addiction is formed. So for me, Valium was employed during the week when I needed my rest in order to be sharp the next day. Come Friday and over the weekend the Valium was not used, but I had to give myself permission not to sleep, if that occurred.

Finally there is another price to be paid with painkillers. Painkillers and most anti-depressants help you sleep but generally do not allow you to get into stage four, deep sleep. Stage four sleep is important for a number of different reasons. A detailed discussion of this is outside the realm of this book, except to note that stage four sleep is associated with dreaming, autonomic nervous system activity and the manufacture of natural killer cells (which fight infections). I often see chronic pain patients who are suffering from colds and flu-like symptoms.

Most of us have had the experience upon waking in the morning feeling just fantastic, ready to tackle the world. This is a sign that you got into stage four sleep. I struggled with this as I woke up feeling okay, but not with a great deal of energy. I would try to do things but it always was an effort due to lack of energy and drive. It was like I was always operating at 60%, and looked for ways to not do things that required energy. Trying to describe this phenomenon is difficult, yet it I is regularly reported by patients on medications. This report sounds amazingly like the symptoms of depression, which it may have been, and I want to make sure that my colleagues are aware of this negative effect of painkillers.

On the weekends when I didn't take the Valium, sleep was variable with some good nights and some not so good. The level of pain affected the quality of sleep, so it was like being between a rock and a hard place. Take the medication and not get stage four sleep or do not take the medication and hope the pain did not wake me up or keep me awake.

Through this carefully designed program I was able to get adequate if not good quality sleep, reduce the pain and avoid becoming addicted. Most physicians are reluctant to prescribe Valium or some such drug so it is important to work with your physician and pharmacist and have a plan in place on how you are going to use — not abuse — medications.

Comments on medications
Remember that pain medications only mask the problem. They do not cure it.

1.   Pain medications are important as they give pain relief. Part of the problem with chronic pain is that the nervous system becomes hypersensitive to the repeated stimulation from whatever is causing the pain. Over time it takes less stimulation to cause more of a reaction. Medications at least dampen this effect, albeit on a temporary basis.

2.   Most pain medications have not been studied for long-term effects. The main regulatory body in the US is the Federal Drug Administration (FDA). The main criteria for inclusion or release of a drug for whatever use is a five to six week trial in a double blind study as to its effects and side effects. The effects in this time period are monitored and studied examining for cure effects and for side effects. The FDA considers this design the gold standard for acceptance and approval. In a double blind study a doctor who does not know what he/she is giving to the subject gives the treatment, usually pills. Also the subject does not know whether they are receiving a sugar pill or the real thing. The results are tabulated and if better than chance improvement is found, the pill is then authorized for utilization. Once this authorization is approved the pill is released to the market place. Once the release occurs the drug can actually be prescribed for any condition, even though it has *not* been subject to the double blind controlled study for that condition. So what happens is medications wind up being used in the market place for pain control or other conditions but have not been subjected to the double blind analysis. Viagra is a good example of this. It was originally approved for the control of hypertension, not erectile dysfunction.

3.   Another issue of concern is that there are no long-term studies on the effects of medication use over significant lengths of time. Remember the FDA only requires five to six weeks of testing. Long-term effects are required to be reported to the drug company who are only required to report to the FDA if the incidence of problems exceeds 5%. However, it is the drug companies who must report this. Also drug companies have in their contracts with researchers that if negative results are found the researcher must report these to the drug company — not the FDA.

4.   The issue of long-term effects of drugs is further troubling. A few years ago a highly respected medical research journal issued a formal written apology to its readers because the journal, in conducting a five-year follow up of the research papers it published, found that 70% of the papers had been disproven. Thus I have taken the position that the

long-term effects of various drugs are just not known and my patients need to know this. I am comfortable with short-term use only. At this point in time I have no answer except to advise people to be aware and to make decisions based on what they learn from the Internet and in consultation with their physician.

5.   The double blind controlled study has become the gold standard for acceptance by the FDA for the control of pain. This is of concern as emerging trends in the chronic pain management fields show studies that combine pain medication in various combinations and/or in combination with active rehabilitation. Active rehabilitation requires the patient's knowledge of the therapy and usually active continued treatments at home. Thus it is impossible for these programs to be evaluated according to the gold standard. The consequence of the FDA position is that only drugs can be studied and recommended. This presents a major block in the regulatory system in finding solutions to the chronic pain problem.

6.   Finally a comment on street drugs. In Canada, use of medical marijuana is legal for certain chronic pain conditions. Several of my patients smoke marijuana, reporting reduction in pain symptoms. Usually the most benefit occurs in pain associated with muscle spasm. Our own clinical research shows that long term and significant marijuana use alters the brain wave patterns seen on the EEG. Otherwise there is just no research that allows me to take an informed position. For example, an article in the British Journal of Psychiatry indicates that excessive marijuana use can induce psychosis in people. This report should be received with caution as it needs to be replicated and/or studied further. Until objective research can clearly define the risks and benefits, I acknowledge the use of the drug but limit myself to that. I do not recommend the use of marijuana. Neither do I condemn its use for the relief of chronic pain. If a patient uses any other street drug, they will be kicked out of the program. I believe their use is harmful on top of the chronic pain and also because they alter the readings that we use in biofeedback.

HOW FAR YOU GO IN LIFE DEPENDS ON YOUR BEING TENDER WITH THE YOUNG, COMPASSIONATE WITH THE AGED, SYMPATHETIC WITH THE STRIVING AND TOLERANT OF THE WEAK AND STRONG. BECAUSE SOMEDAY IN LIFE YOU WILL HAVE BEEN ALL OF THESE.

~ GEORGE WASHINGTON CARVER ~

CHAPTER NINE

Psychologists

~~~~~~~

IT IS ALL IN YOUR HEAD. It was only a minor injury. If you would just try harder you would get better. Quit being a wimp. Quit exaggerating. You are only doing it for sympathy. You are only doing it for the insurance claim. I am sure you have heard it all before.

When the medical professional cannot find anything wrong with you anatomically, they tend to blame the chronic pain sufferer, usually by saying it's in his/her head.

If a doctor prescribes it and it works, it is good medicine.

If another health care provider prescribes it and it works, it is placebo.

If a doctor prescribes it and it doesn't work it is the patient's fault or it is chronic pain.

When the doctor can't find anything physically wrong with you it's your fault and off you go to a psychologist. As in so many of these situations this can be a blessing or a disaster.

When all you have is a hammer, everything is a nail. So it is with psychologists. They by and large are trained only to examine the psychological aspects of chronic pain ignoring the fact that there may be a physiological reason for the maintenance of the pain. With a few exceptions their goal is to help you learn to live with the pain, to try to improve the quality of your life. With relaxation techniques, counseling and other types of therapy the impact of the pain can be reduced.

This philosophy of learning to live with the pain is not one that I wholly support. When my pain first occurred I wanted to get rid of it. I did not want to live with it. As the years have gone by I have learned to live with a little pain, not graciously or happily, but with a recognition that I must.

The patients who come to this clinic do not want to learn how to live with the pain. They want to get rid of it. They have heard that our goal of therapy is to get rid of the pain, with a caveat that they will probably get rid of most of the pain but not all of it.

It is important to recognize that emotions do play a part in chronic pain, but I believe they are never the sole cause of it. Chronic pain sufferers need to make sure their mental health worker shares this philosophy. It is important to understand the role emotions play. I believe that emotions can make chronic pain worse by intensifying, exacerbating and/or maintaining it, and I explain this clearly to my patients. I tell them that medical science does not know everything there is to know about the human body.

As an example, I explain that medical science is just starting to understand what happens to neurotransmitters (biochemicals in the body that transmit the neural signal from one nerve to another) when they are stimulated. Why some neurotransmitters become sensitized to the pain signal (meaning they transmit the pain signal more quickly and efficiently), while some become desensitized (losing their effectiveness to block the pain signal) is a question that medical science is just starting to grapple with. I make sure that all the patients understand that their emotions are probably affecting their pain by increasing the neuro activity of the nervous system making the pain signal easier to transmit and thus perceived as worse.

The goal of psychological therapy is to reduce those factors that are making things worse. I tend to be behavioral and cognitive in my approach, meaning I look for behaviors and ideas that increase a person's anxiety, depression and other emotional factors, and then we work together to weaken or eliminate them. I am clearly not a hand holder, as it is expected that the patient be actively involved in the treatment and responsible for such. I believe that not one therapy works for all patients so multiple therapies are tried in my office.

In general, anxiety and depression are the emotions most commonly seen as involved with chronic pain. These two emotions are usually seen in combination with one who is also tired and wired. The terms anxiety and depression are used in this book in a manner similar to that as used in the pop culture. Anxiety is understood to mean nervous, uptight, unable to relax, physically feeling hyper or strung out or unable to either stop the brain from going around in circles or focusing on one thing over and over. Under the category of anxiety, post traumatic stress disorder (PTSD) is included for individuals that have experienced a traumatic event.

Depression is viewed to include feelings of helplessness, hopelessness, a lack of energy, poor sleep and often weight loss or gain. The main roadblock to improvement is the feeling of hopelessness. Often the patients seen are down and out having tried numerous doctors, therapies and home remedies and have exhausted their financial resources. Some have even lost their marriage. Just trying to get these people to have some hope can be a major challenge.

It is believed that chronic pain involves over activity or hypersensitivity of the nervous system to the pain or other stimuli. So our strategy at the clinic is to bring to bear anything that will reduce the activity of the nervous system. Anxiety is viewed as causing the nervous system to become hyperactive, transmitting the pain signal (and other signals) quickly and efficiently. Thus techniques like biofeedback, relaxation training, EMDR, and cognitive behavioral therapy are used to quiet the system down. Discussion of how the person is coping, where they are in the medical investigation process, and any other aspect of the chronic pain process is conducted. For example most people who come to see me are scared. They are going through a process in which they feel incurable pain. Do they have cancer? Why does it take so long to get into see a specialist? Maybe they are just making the pain up? These are some of the more common issues discussed.

Personally I shared these concerns and thoughts and numerous times used relaxation training and cognitive behavioral techniques to reduce my anxiety level as I went through the numerous investigations. Never once did I think that the basis of the pain was purely psychological, but constantly looked to make sure my behaviors or thoughts were not making it worse.

For example I have a tendency to minimize my pain. Because I viewed the pain as less than it was, I cheated on my rehabilitation exercise programs, skipping a day or two only to pay for it later. Exercise in my case helped reduce the pain, so you would think that I would work out every day. However, like most people, when the pain decreased I would think "Oh, well I'm getting better, I don't need to work out." The hardest part of rehabilitation occurs when a person starts to get better. Appointments are missed, homework is not done, etc, etc, etc. Then when the pain returns they say "Oops," and return back to therapy. I was (am) one of the worst offenders of this pattern of behaviors.

Over the years it became apparent that it was important to differentiate the individual's level of anxiety before the onset of the pain versus what it was during the pain. This is important for people to remember. A person who was highly stressed before the onset of the pain will need more work than someone who was less stressed. The neurological system of a person with high anxiety will be more efficient at transmitting the pain signal.. PTSD really compounds this problem.

Post traumatic stress disorder (PTSD) is a severe anxiety disorder that can develop after exposure to any traumatic event that results in psychological discomfort. The event may involve the threat of death to oneself or to someone else, or to one's own or someone else's physical, sexual, or psychological integrity, overwhelming the individual's ability to cope. PTSD symptoms include re-experiencing the original trauma(s) through flashbacks or nightmares, avoidance of stimuli associated with the trauma, and increased arousal - such as difficulty falling or staying asleep, anger, and hypervigilance. Formal diagnostic criteria require that the symptoms last more than one month and cause significant impairment in social, occupational, or other important areas of functioning.

This pattern is evident in most of the patients who have been in a motor vehicle accident. It usually isn't seen in those patients in whom the pain had an insidious onset with no trauma involved. In PTSD the pain behavior often is reinforced by the continued response to negative external events.. For example a return to the scene of an accident can trigger a massive emotional response that is reinforcing. In working this reaction I believe not only is it important to treat the cognitive and emotional issues, but the physiological response (i.e., increased heart rate, sweaty palms) as well. It is difficult for me to go by the place I slipped even though the restaurant is no longer there. While I know it is gone I still get a moderate fear and anxiety response with increased heart rate and sweaty palms. This physical reaction usually stays with me while I am in that part of the shopping center and disappears as soon as I leave.

Psychophysiological responses can be easily monitored using routine biofeedback procedures. As seen in me, I believe that if the psychophysiological response is not treated it will continue to reinforce the anxiety. This is why purely cognitive or behavioral therapies often fail.

Issues pertaining to depression follow a similar pattern to that of anxiety. Depression is believed to be related to a biochemical imbalance in the body and chronic pain causes the body to burn up serotonin, a biochemical associated with mood and satiety. Depletion of serotonin is hyper arouse the nervous system, making it more sensitive to stimuli, thus increasing the perception of pain. Serotonin depletion also causes a disruption of the sleep cycle producing the tired feeling. In various research studies sleep deprivation has been shown to increase pain perception and also produce fibromyalgia-like symptoms. Thus the chronic pain patient gets caught in a vicious cycle of pain leading to less sleep leading to more pain leading to less sleep and so on. This pattern was very evident to me as sleep became a precious commodity, at times poor and at times good. During the poor phases it was very difficult to cope and feel optimistic and rested. As mentioned, I was lucky to have discovered Valium and its effects on my sleep. Even now as my symptoms gradually decrease over time my sleep remains very important. Some nights I can stay up late, the next night I am in bed early because the amount of pain I experience that day is very tiring.

One specific emotion I want to discuss is anger. As with most of the other things talked about in this book, anger can also be a two edged sword. It is not the anger itself, but the way in which it is manifested. It can serve to motivate, to continue the fight against the pain; however, it can also be rage full and directed towards people.

Often I see patients coming in who are extremely angry, usually at their insurance company or their lawyer. After years of trying to treat these people I now do not take them as patients. I discovered that they won't let go of their anger as they feel cheated or abused by the insurance company, or neglected by their lawyer. Anger maintains the hyperactivity of the nervous system defeating any therapeutic gains. Also some of these patients had incorporated their anger into a belief system figuring the insurance company was conspiring against them. Unfortunately the way insurance companies treat people with claims it gives this appearance.

One of the hardest jobs in the world is to be a receptionist for a medical practice or even worse a chronic pain center. When people come into the center they are hurting; that's why they're there. Some react to hurt with anger and are very demanding and downright rude. Not all patients are like this as some are very sweet and likeable. But it is hard to be happy when you hurt and don't know why you hurt and then wind up in a psychologist's office. As a patient, I tended to bounce back and forth between being polite (as I knew delays happen in the clinic's schedule), and being snippy (not quite rude but somewhat cutting and sarcastic in my comments). Which part of me came out depended on how much pain I was in and how pleasant the receptionist was. On two occasions I remember I was a real jerk, and afterwards felt quite embarrassed.

However there is another side to anger that can be positive. This is the anger that patients display towards their pain and their symptoms. This anger, which I call a good anger, can serve to motivate and to keep motivating the patient to continue to struggle towards getting healthy and taking control of their lives. It basically serves as a motivator to get better. It takes energy to continually go to appointments and function at a reasonable level, while suffering with pain, let alone suffer through any more pain that therapy induces. This type of anger helps to sustain the motivation to get well and not give up.

I would routinely get angry at the pain, the frustration of not being able to do things and the loss of role in MY family. The pain would serve to motivate me to try again, to try different therapies and to continue to challenge the pain. I made the pain the enemy, and projected the pain outside of my body. However in doing this I built a cocoon around myself with denial, anger and minimalization. It was only in hindsight that I realized what I had done, and in doing so pushed all my feelings (including love and caring) out of my space. This of course meant I had pushed my wife, my friends and my colleagues away, becoming isolated in my existence.

My pets filled this void and became my therapists, as I would talk about my pain to them. The cats, but mainly my dogs, would often have a dialogue with me. It is amazing what good listeners they can be, and even better they don't condemn you for what you have to say. Occasionally they would even come up with a good piece of advice, and were rewarded with a milk bone or cat treat.

As my pain started to recede the wall around me diminished and I started to communicate with those closed to me, especially my wife. I realize now that it was a mistake to build the wall around me, but it helped me to survive. As a therapist I do not encourage building a wall around you for you need friends and loved ones to help you cope. However I was powerless to change this, because it was only in hindsight that I realized what had happened. Try keeping a diary and sharing your pain with your significant other; you need that support.

Most psychologists are trained in counseling psychology. The counseling psychologist may be trained in a number of different techniques and procedures dependent upon the school they went to and their personal interests. A few have extensive knowledge of chronic pain, but this field has not yet been well developed by universities or schools. Counseling, relaxation training and various other forms of therapy can reduce the effects of chronic pain. However it is important that the psychologist recognizes that negative psychological factors do not cause chronic pain, they only enhance it. Counseling for chronic pain should be brief and focused, not open ended. It should last only long enough to regulate the nervous system.

A few sessions of marital and family therapy is also recommended as chronic pain forces changes in family roles and dynamics. Couples that have been married for years suddenly find themselves in uncharted waters. The chronic pain sufferer cannot do the things he or she were previously relied upon to do. While reactions to this loss vary from person to person the general reaction is very similar to those of grief: denial, anger, depression and sadness. Meantime the spouse is not sure what is going on and the couple gets caught in a catch 22 situation or paradox. On the one hand should the spouse support and comfort? Or on the other hand does she get angry with the pain sufferer? But they can't get angry with the pain sufferer for this person is in pain, or is he? The doctor says there is nothing physically wrong, but the individual continues to hurt. While this untenable situation may eventually drive loved ones apart it can also make them closer. Understanding the impact of chronic pain on the individual, marriage and family will save many anxious moments, poor communications, and arguments.

I know too well the dilemma the pain sufferer and the family face. One of my pre-pain chores was to cut the grass. It was a chance to get out and exercise, contribute to family functioning and the yard looked good when it was done. This changed the day I could no longer walk while cutting the grass. Try as I may I could not push through the pain. There I was in the middle of the lawn unable to move or bare weight on my right leg in screaming pain. Mary had to help me up to the deck. You can imagine the fear, anxiety and embarrassment that I felt, let alone what was going through Mary's head.

As I lay in bed at home post surgery I would hear Mary out cutting the lawn, on top of everything else she had to do. For whatever reason, the sound of the lawn mower upset me, probably because it represented a total loss of my ability to contribute to the family. Fortunately Mary has a strong Scottish heritage in which she refused to give in to the demands of the family and yard. At times I swear, it nearly killed her and at times selling the house was a serious consideration.

However in the end she survived and acquired several new skills in the process. Eventually we figured out a way we could work as a team with Mary doing the physical things (i.e. climbing up a ladder) while I provided the knowledge as to what to do (i.e. when wiring a new electrical outlet knowing what wires should be connected to what). Even today we share jobs such as cutting the grass, as I cannot do the whole yard as yet. It is important that the members of the family understand the impact of the pain as this can pull it together or drive it apart.

Neuropsychologists are highly trained specialists who use psychological tests to evaluate how the brain is functioning. For example they may use tests of memory to determine if the chronic pain has contributed to memory loss. They are used to determine how a brain is functioning whether it is post stroke or post motor vehicle accident. Insurance companies routinely utilize evaluations of this kind.

Applied psychophysiology is a new field of therapy. This group uses psychophysiological techniques to work towards reducing and eliminating the pain. Psychophysiological techniques are commonly known as biofeedback. "Biofeedback is a process that enables an individual to learn how to change physiological activity for the purposes of improving health and performance. Precise instruments measure physiological activity such as brainwaves, heart function, breathing, muscle activity, and skin temperature. These instruments rapidly and accurately 'feed back' information to the user. The presentation of this information — often in conjunction with changes in thinking, emotions, and behavior — supports desired physiological changes. Over time, these changes can endure without continued use of an instrument." (Definition provided by a joint task force of the Association of Applied Psychophysiology [AAPB], Biofeedback Certification Institute of America [BCIA] and International Society for Neurofeedback and Research [ISNR].)

Psychophysiology is very important from my point of view as it offers a new alternative in the fight with chronic pain. A recent presentation viewed on the Internet suggested that medication by itself is not enough to control chronic pain and an integrated program of medication and active rehabilitation is needed to combat this problem. Medication, counselling, exercise and biofeedback offer such an alternative.

THE HEALTHIEST RESPONSE TO LIFE IS JOY.

~ DEEPAK CHOPRA ~

CHAPTER TEN

The Brain

THE BRAIN IS AN AMAZING ORGAN weighing around three pounds but consuming about 30 to 40% of the body's energy. It is believed that most of this energy is consumed generating electrical signals. When we are younger, there are 100s of billions of cells with trillions of neural connections, but as we age cells die off. Relatively very little is known about the brain, especially as it relates to chronic pain.

Although I thought that I knew how the brain could affect pain, I assumed that this information was not relevant to me. As subsequent events proved, boy was I wrong. Throughout the latter part of 2009 and into early 2010 my pain had dramatically decreased by about 70%. I could walk without aides, sleep with only regular Tylenol, and enjoy life to a large degree. Things like getting out of a chair remained difficult, as did climbing stairs, walking distances and carrying heavy loads. As spring rounded the corner, with the increased activity of gardening, the pain increased, particularly in the left, post-surgery leg. So I started to investigate possible causes of the remaining pain.

Firstly, the muscle activity in the legs was re-assessed using surface electromyographic (SEMG) techniques. Having studied SEMG evaluation in graduate school and subsequently researched and published in this field, I felt this was one area where I was an expert. My SEMG results showed that the muscle activity and firing patterns were normal. I then knew that muscles (while getting stiff from overuse) were not the cause of the pain. So off it was to see the orthopaedic surgeon again. Blood work, X-rays and a bone scan were ordered.

Due to my work schedule these investigations were not conducted until several weeks later. During this time I saw a television presentation by Dr. Daniel Amen on the brain and its effect on the body. . Although familiar with Dr. Amen's work I had never applied it to myself, only to my patients. After listening carefully I conducted a self-diagnosis and I started taking a combination of vitamins and nutritional supplements. I figured, "What the hell, the pain isn't getting any better, and I doubt if it will harm me, so why not?" After a period of about two weeks the pain started going away until today it is almost completely gone.

The results of the X-rays and bone scan came back. All were negative.

So now I am thinking, "How do we account for the pain decrease, if everything physically was negative?" As my pain decreased, my mental functioning became sharper with an increased ability to concentrate and focus for longer periods of time. Energy increased, sleep and overall functioning improved. While it is possible that something physically changed in the time period between when the X-rays were done and the pain decreased, I believe that something happened in my brain. What exactly happened I am not sure, but I do have some thoughts on this.

In 1965, Drs. Ronald Melzack and Patrick Wall proposed the gate control theory of pain, for which they won the Noble Prize in medicine. They proposed that the pain signal is controlled at the dorsal horn in the spine via a gate-like mechanism that allows some signals to get through and yet held back others.

Years later I attended a conference in which Dr. Melzack spoke about the progress in research on pain and chronic pain. At that conference he bemoaned the development of the gate control theory as it directed most of the research in pain control towards the neural networks in the dorsal horn and neglected the role the brain played. Dr. Melzack had postulated that the brain is also critical in our understanding of chronic pain, and too much research was directed towards the dorsal horn theory.

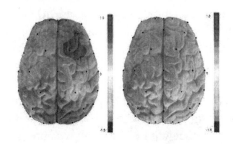

In the clinic I took this to heart and started assessing brain wave activity of the chronic pain patients using a process called quantitative electroencephalography (qEEG). Quantitative Electroencephalography (qEEG) is a procedure that processes the brain's electrical activity from a multi-electrode recording using a computer. This multi-channel EEG data is processed with various algorithms, such as the "Fourier" classically, or in more modern applications "Wavelet" analysis). The digital data is statistically analyzed, and is compared to "normative" database reference values. The processed EEG is commonly converted into color maps of brain functioning called "brain maps." I incorporated these results into a complex treatment plan, which included the peripheral biofeedback modalities, physical treatments and cognitive therapy and began training the brain. The patients' success rates went up, providing pain relief for the majority of them.

While integrating EEG neurotherapy (biofeedback) within my treatment plans I discovered neuroplasticity. Neuroplasticity (also referred to as brain plasticity, cortical plasticity or cortical re-mapping) is the changing of neurons, the organization and re-organization of their networks and functions via new experiences. According to the theory of neuroplasticity, thinking, learning, and acting actually change both the brain's physical structure (anatomy) and functional organization (physiology) from top to bottom.

In his book "The Brain That Changes Itself," Dr. Norman Doidge discusses this concept, particularly as it relates to various types of pathophysiology, including chronic pain. Having been in pain for years and without realizing it, I learned that my brain had changed to accommodate the pain signal! Since the discovery of this phenomenon numerous supporting research articles have emerged. Many of the articles discuss chronic low back pain, whiplash, and closed head injury.

The research literature for amputation surgery also indicates that a spinal block given before a general anaesthetic reduces or eliminates phantom pain, whereas just giving a general anaesthetic does not provide the same results. This suggests that the brain becomes entrained to the pain and the spinal block prevents this entrainment.

I also read two books by Dr. Amen called "Change Your Brain Change Your Life" and "Change Your Brain Change Your Body." The latter book that was inspiring to me. As noted the change was dramatic.

So the questions I pondered were: 1) "given the fact that the orthopaedic and soft tissue causes of the pain had been reduced or eliminated was the remaining amount of pain I perceived due to changes in the brain, and 2) was the decreased level of pain due to the nutritional supplements I was (am) taking along with the brain exercises I had started?" The scientist in me sits and stares at these questions in disbelief. It is much easier for me to believe this as it applies to someone else, i.e. my patients rather than to me.

I have struggled with the notion of the brain's involvement in the cause and maintenance of chronic pain for several years. Although it may be by chance, our chronic pain patients all appear to have some sort of brain dysfunction. Some are post trauma (i.e., motor vehicle accident) or post sports injury, usually one to two years post accident. The qEEG found that some have abnormal (too much or too little) electrical activity or a disturbance in the ability of different parts of the brain to communicate.

Many who have been involved in a car or sports-related accident sustain a closed head injury. A closed head injury involves one or any of the following: 1) a blow to the head, 2) a sudden deceleration of the body, such as whiplash in a motor vehicle accident or 3) a compression injury to the neurons, often due to an explosion. In the first two cases the brain hits the inside of the skull bouncing back and forth (commonly known as a coupé – contra coupé injury) injuring neurons. In a compression case the force of the impact compresses the neurons, killing or injuring them. This injury to the brain can be associated with a minor accident where there is less jostling.

For years we have seen chronic pain patients who responded poorly to physical therapies not only at this clinic but also at several others. Over time it was realized that treating the brain was as important as treating the pain. Often the brain-injured patient will report that when they start a new physical therapy they get about a 30% improvement, and then fail to improve any further. In our clinic, we now treat the brain first followed by physical treatments.

Needless to say I am an example of treatments that partly worked. I received myofascial release, SEMG neuromuscular retraining and physiotherapy from other clinics with some relief usually temporary in nature. In thinking back on my accident I remember the fog that I was in immediately after the fall and wonder if I didn't sustain a closed head injury from the accident. Unfortunately I didn't have a qEEG done. The thinking in 2001 was different than it is now. Today I would not be admitted into the clinic's program without being provided a qEEG.

   As I read this I say "Thank you, God" for it is so nice to be able to sit here and not hurt or be fatigued or grouchy or any of the many things that go with chronic pain. I am still a little cautious for it has only been a few months since I have been virtually pain free, but gee it is nice. Now if I could only learn not to overdo it.

LIFE IS LIKE A TEN-SPEED BICYCLE.
MOST OF US HAVE GEARS WE NEVER USE.
~ CHARLES SCHULZ ~

CHAPTER ELEVEN

Alternative Treatments

~~~~~~~

I wander with some slight trepidation in to this dark but clean clinic in which English is not the primary language. I am here because I hope that Ancient Chinese Medicine will help my pain and blood disorder. The forms take a while to fill out for they are lengthy and cover my medical history in depth. This is somewhat reassuring, but I am still apprehensive.

An individual comes out and escorts me into an examination room. After a few minutes two doctors and a student enter the room. They graciously welcome me to the clinic and start discussing with me why I am there. We spend about an hour discussing things including a detailed review of my medical history and my family's medical history. This latter part of the discussion surprised me for no one had previously done this with me.

They examined my tongue, fingernails and hair, had me lay down on an exam table and poked me in my abdomen. After a few minutes of discussion between them in Chinese they told me what they thought was wrong and how I should go about fixing it. This included acupuncture and a few different types of herbal pills.

After paying a minimal fee I left the clinic feeling quite respected and valued as a human being. I wondered how they could afford to survive on the amount that I paid. There was no comparing this experience to that of the orthopedic surgeon: I would go back to this clinic, but hell could freeze over before I would go back to that surgeon's office. Human dignity is too important to me.

Incidentally the Chinese treatment helped a little with my blood problems but did not help with the pain.

Alternative approaches to the treatment of chronic pain are difficult to evaluate due to the lack of research in these areas. Because of the restrictions placed by various government agencies on claims that can be made, the exact effects and outcomes of clinical trials cannot be advertised.

For example, claims to cure back pain cannot be made so the advertisement has to couch what it says in vague general terms such as relaxes muscles in the back. Thus, as a therapist I am left wondering how to use a treatment.

The field is contaminated with wildly excessive claims of successful cures, which kills the credibility of those involved in pain research. The same issue applies here that applies to drugs. Chronic pain is a complex issue and one type of treatment will not successfully cure all chronic pain.

Because I did not partake in all of the alternative approaches for pain, it was difficult to evaluate them for this book. By nature I am a fairly conservative person and am quite restrictive as to whom I will let treat me. This is partly because I know how most treatment techniques depend upon the skill of the practitioner, and partly because I value my privacy. Essentially I am a shy person.

It costs millions of dollars to take a drug from design through testing to market. As previously stated the gold standard for getting a drug approved is replicated random double blind studies. These same criteria are also applied to alternative treatments thus limiting product development to only those companies that have deep pockets.

As a company cannot make claims about "cures" for specific conditions without this testing, they get around it by making non-specific claims like "helps promote healing" or "promotes relaxation." This leaves practitioners scratching their heads as to what the product or treatment specifically does.

Drug companies fund most research. They naturally fund only those products they are developing. So research funding for alternative treatments have to come from other sources. A potential source of revenue for research monies is research foundations.

Research foundations tend to adhere to areas of research that can be proven with the double blind criteria and the scientists involved tend to be very narrow in focus, concentrating on microscopic aspects of a particular problem rather than anything applied.

Years ago I was a finalist in a competition supported by a foundation. The proposal was to further study my doctoral work by studying muscle activity in chronic back pain and learn how a treatment technique would eliminate or decrease muscle pain. This is an area known as applied research. The other finalist was studying the effects on the biochemical properties of a knee ligament by pulling on a leg 1,000 times. The subjects were rabbits. The rabbit research won.

Thus to give the reader any sort of reasonable advice this chapter is divided into two parts: a) what I learned as I went through the years of pain and attempts at relief and b) what I have learned from my patients.

The first thing that I learned was that everyone has a "cure." If everybody has a cure why are so many people still hurting? I used a couple of criterions to judge whether to try something.

1.   What are the claims and are they physiologically sound?

2.   Can they hurt or harm me; what is the risk involved?

3.   How long was I going to need to take or do something before the "cure" would take effect?

4.   What are the qualifications of the person working on me?

I tried physical therapy several times. Most of the time the relief was short lasting, usually two to three days. While I did experience some pain relief, once or twice I was left in severe pain. Remember the trip to emergency? I learned that physical therapy outcome varies upon the degree of postgraduate training the person has, and how much experience they have with chronic pain. It was always in the back of my mind that the treatment should not irritate or flare the nervous system except for a brief period of time. I learned not to go to a jock doc for chronic pain because their techniques are too rough and aggressive (no pain, no gain) causing further irritation of the nervous system.

Massage therapy was used regularly and produced significant pain relief, but again short lived. The same principles as for physical therapy, massage therapists are primarily trained to treat sports injuries. Massage for chronic pain requires a much gentler touch with some lymphatic drainage. Massage therapy occurred weekly during the time period before surgery and helped me need fewer drugs.

Exercise, gentle as possible was tried earlier in the time period when the pain first occurred. Even while on crutches I experienced some relief, as it felt good just to get up and move around. It is important to keep your blood circulating through activity and not remain sedentary.

Biofeedback training, particularly using SEMG techniques helped reduce the pain in my muscles but not in the joints. This worked by teaching me to focus on the various muscles in my legs to make sure they were working correctly and to relax them when they were in spasm.

About once a week I get asked to try, and hopefully, recommend a new form of treatment, herbal drug, or other healing product. Some I try, some I do not. I review the product information, Google it and then maybe try it, depending upon the risks and benefits. This is what I recommend:

1.   Read carefully and thoroughly any written materials that you find on what you want to do or try.

2.   Recognize that the more specific the claim the better you will know the benefits and risks. Do not believe that one product will cure everything just like one medication will not fix your chronic pain.

3.   Google the product or procedure and read all you can. The Internet is a great source of information.

4.   See if you can find other people that use the product or procedure.

5. Don't commit to years of treatment or years of drug use. If something is going to work it will usually work in the first six weeks.

6. In the end, recognize that not every product works for everybody. Do what you feel is right for you, not what is right because somebody told you to do it.

I use biofeedback procedures to treat most of my patients. Some of my colleagues and medical doctors consider this "flakey." I use this form of treatment because there is enough research (more than on most drugs) to show biofeedback is safe with little known side effects or risk to the patient. Yet biofeedback treatments are criticized because there is not enough double blind research on these techniques. "Hogwash" is my response! In 1999 I was part of a team that compiled a list of research articles on SEMG. We found 2,600 research articles. Since then, another 6,000 plus articles have been published.

Additionally, there are over 90,000 articles published on qEEG, using this technique to aide in the assessment of various disorders. So I feel quite comfortable in using these techniques and recognize that the criticism is political and not scientific in nature.

Also I highly recommend the use of occupational therapists. It isn't so much a matter of them curing the pain as much as they teach very practical procedures to reduce the impact of the pain and associated movements.

Saying anything about chiropractic treatments is difficult, because the patient's feedback at the clinic about this field is so mixed. On the one hand some people swear by it, stating it reduced their pain and "saved their life." On the other hand I have seen people who were injured by their chiropractor.

There is an extensive volume of research about chiropractic treatment as well. In general it points to positive outcomes and improved health. However the propaganda put out against it, usually by the medical profession, is that there are no double blind studies to support its efficacy. Sound familiar? The criticism, as with that for biofeedback, appears to be more for political purposes than is scientifically based. However this serves to create confusion for the patient and their health care providers.

One form of chiropractic treatment that I am supportive of is the NUCCA protocol (NUCCA = National Upper Cervical Chiropractic Association). This is an advanced training that chiropractors can take. It focuses only on the upper neck joints (C1, C2), which are particularly important when someone has experienced a whiplash accident in which the head was rotated and/or snapped.

Other forms of treatment such as naturopathic medicine, homeopathic medicine among others have logical theoretical foundations. I just don't know enough about them to comment on them so would encourage individuals who are using them or planning to use them, to research the area well and continue to investigate their claims.

Finances also limit the use of alternative health care techniques. Insurance coverage, if at all, is usually very limited. During the course of the pain I spent over $10,000.00 dollars attending appointments, with little reimbursement. It is important to remember that insurance coverage is based on the acute care medical model not on treating chronic conditions. Also, don't count on it to help with payments for alternative health care.

I wish that I could tell you to take this or that for your chronic pain. I can't because there just isn't enough research on alternative choices. The medical and insurance systems are set up to reinforce drug use and inadvertently (or otherwise) restrict research in other fields. With the changes that are coming in the field of chronic pain, driven by the knowledge that medications by themselves will not solve the problem, it is hoped that comprehensive studies involving medications and alternative forms of treatment will help to reduce or solve this painful problem.

THINGS WHICH MATTER MOST MUST NEVER BE
AT THE MERCY OF THINGS THAT MATTER LEAST.

~ JOHANN WOLFGANG VON GOETHE ~

CHAPTER TWELVE

Reducing The Pain - Taking Control of Your Life

~~~~~~~

Over the last year I have come to realize that I will have to live with some degree of chronic pain for the rest of my life. The level of the amount of pain will vary, dependent upon a number of factors. Some factors like the weather are outside my control, while other factors such as mood, stress and physical activity are well within my ability to regulate.

My purpose in writing this book was to share with you from both a professional and personal level information about chronic pain and how to beat it. As I continue along the voyage I want you to hear this message: until there are more advances in medical science and changes in health care delivery systems you are facing a tough battle. You can let the chronic pain control your life, or you can do things to reduce (not cure) it.

If you want to let the chronic pain control your life do the following: believe that taking only medication(s) and not doing anything more will cure the pain. Take a pill, usually a painkiller or an anti-depressant, focus on your pain, sit on a couch (don't exercise), isolate yourself from friends and family, and turn your brain into mush by watching too much TV. See your doctor intermittently, whine to him/her and get more pills, preferably stronger and more addictive. Complain to any and everyone what a rotten deal life has given you and how hopeless your case is. Congratulations, you have now become a chronic pain in the ass, as well as a chronic pain patient.

Or you can do things to control pain and start to move towards health. Through my adventures in trying to get rid of the pain I have learned a few things. The following materials are broken up into two sections: a) things that you can do that apply to most forms of chronic pain and then b) brief pointers concerning specific things that you can do for specific problems.
Things to do For Most Chronic Pain Problems.

Before delving into the details of what to do, I want to explain to you what I believe causes chronic pain and what happens to the neurological system. Chronic pain in most cases involves the central neurological system, which includes the brain, the spine (dorsal horn) and neural transmitters.

The nervous system becomes sensitized to the pain signal. What this means is that it takes less of a pain signal to cause more of a pain perception. This is known medically as hyperalgesia.

There is also evidence that non-pain nerves (i.e., mechanical receptor nerves) can change to become sensitive to the pain signals and add to the volume of signals that impact upon the brain. This is known as allodynia.

At the nerve junction the biochemicals become sensitive to the pain signal transmitting the painful stimuli across the nerve synapse quicker with more efficiency. The brain reacts to the repeated stimulation by becoming more efficient at transmitting the pain signal particularly to the part of the brain that is known as the homunculus.

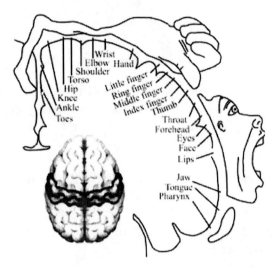

A schematic of the homunculus based upon the area devoted to specific motor areas of the body

The motor homunculus is a pictorial representation of the body sitting in the middle part of the brain stretching from just above your ear to just above your ear on the other side. Each part of the body is represented by an area in the sensory motor strip (homunculus).

The homunculus has been shown to react to changes in the sensory information that comes to the brain. If stimulation to the brain is decreased then the size of the area in the brain that should be stimulated decreases. Increased stimulation (pain) is thought to increase the size and receptivity of that area of the brain.

Eventually the brain changes physiologically and eventually structurally (remember neuroplasticity?). These changes can produce a pain signal without any stimulation from the peripheral nervous system as seen in phantom limb pain. Any other neural signal can also add to the stimulation (i.e., emotions cause increased neural activity). Much like the way we learn to play the piano, the brain and nervous system learn to play the pain signal. Thus knowing the length of time a person has been in chronic pain is very important.

So the overall treatment strategy used in treating chronic pain is based around the following questions: a) how do we reduce the activity of the nervous system without irritating it and b) how do we change the patterns of activity so they return to a more healthy state?

IN GENERAL, WHAT CAN YOU DO FOR CHRONIC PAIN?

THE BRAIN:
As previously mentioned, the brain is one of the least studied aspects of chronic pain. I routinely am absolutely amazed when I read the agendas for pain conferences and there isn't even one seminar on the brain and its role in chronic pain. Given that the pain pathways and the organs (i.e., insular cortex) involved in acute pain are well established and known, the lack of presentation about this just blows my mind. Additionally, whether a brain is healthy or unhealthy (i.e., injured in an accident, damaged by drug or alcohol abuse, etc.) is vitally important in the assessment and treatment of chronic pain.

All patients who come into the clinic for the chronic pain program are evaluated with the quantitative electroencelphalogy (qEEG). Experience suggests that individuals with a healthy brain respond to physical treatment much quicker than individuals that don't. Chronic pain patients with unhealthy brains are much more difficult to treat and take longer to start to show progress.

Dysfunction in the brain may be caused by several different factors. Some of these include closed head injury, viral infections, anesthetics, drugs and chronic pain in other parts of the body. Research in these areas is scarce and limited at best,. Thus the material that follows is based upon Stu's experiences and Dr. D's clinical and research experience in treating chronic pain and several other conditions such as attention deficit hyperactive disorder (ADHD) and closed head injury.

Keep your brain active:
I can't emphasize this enough, for keeping the brain active helps combat depression, helps maintain mental sharpness and serves as a distraction from the pain. The old adage "use it or lose it" applies here. It is evident that activity promotes different parts of the brain to remain healthy and lessens the impact of the chronic pain signal on the brain. Brain exercises that you can do to reduce the impact of the pain have also been shown to combat the effects of aging. Try doing a familiar routine in a different way. For example brush your teeth with the opposite hand. This changes the pattern of neuron firing involving more and different neurons causing more blood flow to the brain.

Try crossword puzzles, word search, anything that involves concentration and focus. I have discovered a passion for Mahjong and Sudoku. There are hundreds of games on the Internet, just remember to chose only those that require focus and concentration and regularly try new games. Games of violence don't qualify.

Try doing things you haven't done before. My wife Mary is learning how to speak Spanish. Writing this book is a consequence of me trying to do new things; it requires focus and concentration and has the added benefit of distracting me from my pain.

Watching TV is not being active as watching most TV programs is a nonparticipatory, passive receptive activity.

Feed your brain well:

I believed that I eat a balanced diet and get adequate nutrition to the brain. However, the reaction to taking nutritional supplements was remarkable, so I strongly recommend that every person suffering from chronic pain take a good look at their diet by reading Dr. Amen's books and/ or work with a dietitian to get your brain healthy.

Research shows the constant bombardment of the brain by pain signals depletes the brain of neurotransmitters and other biochemicals that need to be constantly replenished.
Emerging research indicates that North Americans don't get enough of Omega 3 in our diet. I have become a strong supporter of the use of Omega 3, vitamin D and a multi-vitamin complex and encourage my patients to use these and/or other supplements after consulting with their doctor.

Be honest with yourself:

Being honest with yourself means accepting the limitations that your pain places upon you. There are certain activities that you will be able to do, and certain ones that cause too much pain. This is a two edged sword as you do not want to do activities that will increase your pain, but don't sit around doing nothing either.

I like to encourage people to take stock (make a list) of how they have changed, for the better or the worse, then encourage them to select those parts of the list that they want to change and those parts that they feel they can live with. Changing the whole list is not an option. What this serves to do is to get people to focus on what is important and what they can let go of.

Acceptance of limitations is very important because it helps people become at peace with their pain. I realize that I will never be as good or return to the level of physical functioning that I was at before the accident. This is a hard realization to accept, but reality is reality. I would like to be able to golf again, but I can't walk that far, so I have to find something else to do. By admitting this to myself I reduced my feelings of depression, hopelessness and frustration thus reducing the negative emotional input to the neural system. Also by accepting my limitations I don't force myself beyond my physical limitations, thus preventing a pain flare up and its consequent increase in neural activity.

Take Time To Heal:

This sounds crazy because everyone wants to heal, but this needs to be translated into behaviors. Plan your day so you get time to relax, decompress and just chill out. For me fortunately Mary was understanding of this and gave me lots of space to be by myself. I would do deep breathing, have a quiet drink, daydream, and go for a walk: whatever seemed good for me at that time. I felt I was being selfish, but I also knew how important this time was in order to quiet down the nervous system.

Take 10 minutes to do what relaxes you.

Organize to do gentle stretches and exercises.

Cut down on appointments by organizing them so they all don't occur on the same day.

Give yourself permission to be in pain.

Emotions:
Anger is a natural consequence of being injured, such as in a car accident, especially if it wasn't your fault. Give yourself permission to be angry but direct it to the pain not the insurance company.

Anxiety is a natural reaction when the medical profession can't find anything wrong with you. We are so inundated with messages about cancer and other horrible diseases that it is only natural to wonder if that is what is wrong with you. Seek appropriate medical support.

Any significant emotional reaction will need to be dealt with because any added stress increases pain. Decreasing the activity of the neurological system is one of the goals of therapy. This can be done with numerous techniques including but not limited to: talk therapy, eye movement and desensitization and reprocessing therapy (EMDR), hypnosis, biofeedback and relaxation training.

I use the latter two techniques quite extensively, preferring autogenic relaxation training, deep breathing techniques and passive visualizations to reduce the neural activity. EEG neurotherapy is used extensively in this clinic to rehabilitate the brain (if injured) and to reduce brain wave activity associated with excessive emotions. EMDR is highly thought of and is recommended if done by a skilled therapist.

EXERCISE:
Any physical activity that increases blood flow to the brain is recommended. Twenty to thirty minutes of activity 3 to 4 times a week get blood to the brain. Research by Dr. Amen shows that blood flow to the prefrontal cortex is reduced in ADD, while increasing the blood flow helps reverse the signs of ADD.

Muscles:
Pain from a muscle (known as a trigger point) can be excruciating. Ask anyone who has had a heart attack. When compared to pain activation from a tear in the skin, pain activation from a muscle (trigger point) has been shown to be 90 times stronger and last 90 times longer at the dorsal horn (the place in the vertebrae at which the pain signal from all sources arrives). The pain can be so intense that it causes visceral responses such as nausea, sweating and even fainting.

I have studied and researched muscle activity and trigger points for over 25 years. I am amazed by the lack of knowledge by health care professionals concerning this vital part of the body. The role of muscles in chronic pain is poorly understood. This is because medical science doesn't understand that: a) muscles work in units of at least two (usually 4 to 8 muscles), b) that the pain from muscle dysfunction shows up in a different location than where the trigger point is, and c) current drug regimes try to mask the pain rather than fix the problem.

Trigger points are areas of electrical hyperactivity that are sore to touch and cause referred pain. Trigger points are classified into two categories: latent and active. The spot that is sore to touch is considered latent, while an active trigger point refers pain much like a heart attack will refer pain into the left arm. For example trigger points in the neck will refer pain into the head, causing headaches. At various times I have had trigger point pain in my low back, buttocks and legs dependent upon what I was doing.

Pain from trigger points develops in muscles that have been exposed to trauma, chilling, fatigue and acute overload. Generally trigger points take about six weeks to develop so there should be a pattern of decreasing pain (as the swelling goes down) to be followed by increasing pain due to trigger point activity.

Trauma, such as a motor vehicle accident, can cause damage by direct blunt force (such as a blow) or by stretching the muscle along its length while contracting. I see many patients with headaches and neck pain which is often seen post motor vehicle accident following a whiplash in which the head is rotated against its will. While no lacerations may be seen, the interactions amongst the muscles in the neck are disturbed; producing trigger points causing headaches and neck pain. Doctors say there is nothing wrong because they are using the wrong method of assessment. Without proper treatment this problem could go on for the rest of a person's life. This is what the insurance industry refers to as minor soft tissue injury.

Rehabilitation specialists treat the pain, often failing to recognize the muscles that aren't working properly, concentrating on the spot that is sore. Heat and stretching can reduce the pain but until the dysfunctional muscle is fixed the problem will continue. That is why with rehabilitation therapy relief is often only temporary. Presently through the use of SEMG techniques the dysfunctional patterns can be observed and the dysfunctional muscle(s) can be retrained, eliminating the cause of the problem.

Chilling causes reduced blood flow to the muscles producing a buildup of lactic acid and stiffness. This happens because the blood vessels in the arms and legs get smaller (vasoconstriction) forcing the blood to the core of the body. Thus oxygen to the muscles is decreased and lactic acid builds up. Any sort of caffeine (coffee, coke) produces the same effect, thus I limit my patients to one cup of coffee a day in an effort to promote good blood flow to the muscles. I can attest to the effects of chilling as the muscles in my hips stiffen up the moment the temperature gets to below 10 degrees Celsius. I am lobbying for a move to Hawaii, but somehow I don't think an insurance company will pay for it. A glass of red wine also helps with blood flow to both muscles and the brain. However the main issue for me is the vasoconstriction caused by temperature and how it cripples me.

Fatigue is associated with chronic repetitive movements especially when a muscle is overloaded due to posturally incorrect positions. One of the main culprits for this is the use of computers in daily living. A number of people are regularly seen with shoulder pain produced by incorrect positioning while using the computer.

Often people thrust their heads forward while looking at the screen (making one look like a turkey). This throws the center of gravity off and makes the neck and shoulder stabilizers overwork.

I constantly am reminding my patients to "tuck their chins in." Medical professionals generally look only at one muscle at a time, looking for lacerations, bruising and swelling. In general these go away in six weeks and with resolution, it is concluded that there is no dysfunction. What is missed in this assessment, however is that the interaction of the injured muscle with all the other muscles is still disregulated. Muscles work in what are known as myotatic units. A myotatic unit is all the muscles that produce a single movement. Different muscles combine in different ways to produce different movements. One muscle (the primary mover) has to contract, while its antagonist (muscle that opposes it) has to relax, while the other members of the myotatic unit (known as synergists) support, assist and enhance the activity. So to effectively evaluate myotatic muscle activity, the muscle and its partners have to be evaluated during movement. SEMG is the best to do this.

Imagine taking a spring and forcing it to stretch it to its longest length. The spring eventually loses its tension and does not work as a spring after that. When muscles are stretched upon their length while contracting, its neurological signal to its partner muscles and to the brain is altered. Usually this alteration takes the form of a decrease in the muscle's ability to generate inhibitory signals, meaning other muscles in the myotatic unit become hyperactive and overwork to compensate for the stretched muscle. This causes acute overload in these muscles, eventually creating trigger points and pain. I explain to my patients "imagine working with a partner who doesn't do his job and you wind up with all the work. Guess who screams?"

I am very proud to have won an International award for my research in this area.

Often patients will report that if they do nothing they don't get pain. As soon as they do something they get muscle pain. That is because the activity overworks the overloaded muscle and flares the trigger points. People are between a rock and a hard place as it is important that they maintain muscle tone so the muscles can continue to be able to function efficiently. Research indicates that a person lying sedentary loses 3% of the muscle tone per day, so we need to stay active. However if they overdo it they will flare trigger points, increasing the pain and irritating the nervous system. So basically the muscle needs to be stretched (to help it relax) and strengthened (to help it carry more of a load). The question is what to do when placed in this rock and a hard place situation?

Keep as active as possible.

Exercise only to tolerance.

Get into a routine of gentle stretches daily preferably two to three times a day.

When exercising keep the muscles warm, and stretch before and after the workout.

Correct poor posture with guidance from a health care professional. It is important to keep your body centered.

If you see a therapist make sure he or she assess all the muscles around a joint, because finding the weak one is more important in the long run than reducing pain in the sore one.

Talk to your therapist about using heat and ice.

Generally heat helps the muscle relax and get rid of the lactic acid, while cold is used to reduce swelling. Hot tubs are great for relaxing stressed muscles and joints.

If you flare something up - go easy, as you want to reduce the nervous system irritation.

In 1992, the year the summer Olympics were in Spain, I worked with some of the athletes. Their motto was "No pain, no gain, no Spain." Conversely the chronic pain patients' motto is "Pain means no gain and an irritated brain."

Gentle exercise:
Gentle, gentle exercise will pay larger dividends in the long run as this makes up for a loss of muscle tone. Yoga is particularly recommended.

Remain flexible and open:
While right in the middle of writing the section on trigger points and pain I tore a fibre in a muscle (iliocostalis lumborum) in my back. Don't ask how I did it. Except note that it is suggested that stupidity runs in my family. It is hoped that this experience points out how a diverse and timely treatment plan is necessary to deal with pain caused by trigger points.

The initial response was severe pain in the low back that wrapped around into the abdomen. The pain was so intense it caused nausea. Painkillers and lying still worked for a few hours, but sleep was pitiful that night. The next day the pain was so intense I had to cancel appointments and go home to the painkillers and lying still. As I tend to be a restless sleeper all of my movements in bed were irritating. Out came the major tranquillizers so I could sleep without moving, leading to reduced pain in the morning. Any sudden or excessive movements set off the pain and nausea again so movement became very guarded and restricted during that next day. Over the course of the first week the pain slowly started to subside so a program of stretching was initiated. During the second week there was still pain and some slight inflammation. Heat was applied to the sore muscle further improving the situation. The pain was still present over a period of months but massage therapy, stretching and heat were slowly helping me heal. Because I worked with the whole muscle group, rather than just the source of pain, I improved. I slowly stopped taking painkillers and began strengthening my back muscles.

 *DRUGS:*
As stated I am not a physician, I am a psychologist. The materials that follow are based upon my clinical experiences as a psychologist and are not intended to replace advice from a qualified physician. Always do what your physician advises, and if it isn't working you need to discuss this with him/her first.

When in acute pain use whatever drug(s) the doctor has prescribed for you, exactly as prescribed. You want to get rid of, or at least reduce, the pain so you can reduce the impact on the nervous system.

When seeing your doctor always make a list of all medications that you are presently taking. Drug interactions can exacerbate chronic pain problems. After seeing your doctor, if you are still uncertain get a consultation with your pharmacist.

Research the medication(s) that you are on: Google them on the Internet. In the year prior to my first surgery I had so much muscle pain I could hardly walk. I was taking a blood pressure medication, which I learned had a possible side effect of increasing muscle pain. When informed, my physician changed my medication. Within two weeks the muscle pain decreased by 30%.

If your doctor has such a big ego that this type of help is rejected, find yourself another doctor.

As a society we are used to thinking of drugs in a manner similar to that of antibiotics in which the medication solves the problem. Remember painkillers are just that, they are designed to kill the pain, not resolve it.

Remember painkillers and anti-depressants can be habit forming, can be addictive and can ultimately make your pain worse. I routinely see patients who have been on the same medication for years, with the dosage being gradually increased as the pain killing effect wears off.

Most painkillers cause a rebound effect when you come off them. The rebound effect is the tendency of some medications, in sudden discontinuation, to cause a return of the symptoms it relieved, and that the symptoms are stronger than they were before treatment first began. This issue has to be dealt with before I treat the pain; otherwise I have no idea how much of the pain is due to the use of the medication and how much is due to actual physical problems. For example, years ago I saw a woman from another part of the world that was in severe pain. She was taking six different types of painkillers and other medications. After she detoxified (most of the medications had been withdrawn) her pain level was 60% less than when she first came in to the clinic.

Medications complicate my work because they change physiological measures including electromyograph. Additionally there is little published research available on the effects of medications on brain wave patterns. Given that the brain is involved in the chronic pain cycle and that we incorporate braintraining, when one is medicated it becomes difficult to know what treatment to provide. So it is preferred that the patients are off of pain medications.
Remember long-term effects of most medications are not known. I personally have only taken blood pressure medication in the long-term. Painkillers should not be taken for extended periods of time and if they have to be then different types should be used for predefined periods.

Antidepressants make you feel good. They do not cure the pain. My opinion is that the main advantage in using an antidepressant is that it should give you the energy to work on your rehabilitation (i.e., exercise). However most people take these to feel good and continue to sit on their butts doing nothing. It is really important that people use the anti-depressant to get going. One of my doctors gave me great advice: If you are going to do something that you know will flare the pain, take the pain killer before you start, not after.

It works.

In conclusion, drugs are a two edged sword, a blessing when you are in acute pain, but also potentially addicting and can make chronic pain worse. Use them judiciously and only under appropriate medical supervision.

One issue that is of grave concern is that of caffeine. Patients who drink more than one cup of coffee, or for that matter anything with caffeine in it except Chi tea and green tea, per day are refused entry into the chronic pain program. Caffeine reduces blood flow to the muscles and alters brain wave activity. Muscles and their trigger points need all the blood they can get, so don't reduce it, unless you want to make the pain worse. Caffeine increases brainwave activity in the front of the brain giving people an energy boost masking the effects of fatigue. The new energy drinks have even more caffeine in them making them more of a concern.

Relaxation training is a critical part of any program. While I prefer autogenic training (a form of self verbalization using mantras focusing on different parts of the body) there are a number of different types of relaxation tapes and programs available. I encourage my patients to try some and find the one that works for them. It is strongly suggested that patients dedicate 20 – 30 minutes a day for this exercise. It seems a lot of time but when in chronic pain, wisely use your downtime to relax.

HELPFUL HINTS FOR REDUCING THE
PAIN OF SPECIFIC CHRONIC PAIN PROBLEMS

Over the years I have encountered a number of procedures that help to reduce pain for specific problems. These suggestions are designed to reduce the pain but are not intended to replace recommended treatments from licensed therapists. Please try them carefully.

HEADACHES
When medical investigations can't find a cause for your headaches, get the neck muscles (sternomastoids & cervical paraspinals) and/or jaw (masseters & temporalis) checked out for trigger points.

Use ice on the head where it is sore.

Use heat and stretching on the muscles in the neck and jaw that are sore.
Daily relaxation training using whatever tape works for you. This is one chronic problem that stress seems to be a big part of.

Medications can cause severe rebound effects particularly with headaches. See your doctor and work with him/her on gradually reducing the amount of medication you are taking. This will initially increase your headache but in the long run should decrease frequency and intensity. Be sure you are doing your stretches, getting massage therapy and using relaxation tapes while you are trying to withdraw.

If the headache appears post trauma make sure you don't have a concussion. Blood from mild injuries usually takes about six or seven days to pool sufficiently to show up on an MRI or CT scan.

So if you can get one of these done immediately post trauma you should talk to your doctor about getting one done a week later just to be sure there is no internal injury.

If you are post trauma see a physical therapist or chiropractor that specializes in the alignment of the spine at C1/C2 (two top vertebrae of the spine).
If you have a history of concussion remember that the effects of the next one will be worse because the effects appear to be cumulative. Snowboarders remember to use the board and not your brain to surf on.

LOW BACK PAIN

Research shows that 80% of back pain comes from muscle spasm (trigger points).
For acute pain use cold (i.e., ice, bag of frozen peas) for up to two days (do not leave the cold on continuously as it can cause frostbite). Then use heat afterwards. Cold reduces inflammation and heat relaxes the muscles. When heating try gentle stretching but don't overdo it. If you flare the pain use cold again and start all over.

Muscle relaxants are recommended over painkillers for the acute pain stage. Talk to your doctor about this.

Stretching for the back itself will depend on what feels good. At the clinic our treatment includes bending (either forward or backwards) just to the point of pain and hanging out there while doing deep breathing. I use a device I invented to support the patient while doing flexion and extension so the patient can relax and focus on the stretching.
Stretching all muscles involved is important. Muscles involved with back pain go from the knees up to your shoulder blades. Particularly make sure you stretch your hamstrings (muscle in the back of the leg/thigh), and gluteal muscles (you sit on them) as these muscles affect the stability of your pelvis.

Use the handicapped stall in washrooms. The seats are about 6 inches higher than a regular stall. This elevation reduces the degree of flexion, reducing unnecessary stress to muscles.
Similarly get a plastic donut that will raise your toilet seat at home.
Use an extended shoehorn to put your shoes on. Get some slip on loafers and use the extended shoehorn to put them on. This saves having to bend over.

As soon as possible go for short walks over uneven ground. Do this to the point of pain. Uneven ground is recommended, as this will force some of the smaller muscles to work.
Swimming is also highly recommended. I can swim to the bottom of a pool quite well: I just have trouble getting back up to the surface.

Don't sit for long periods of time. This stretches the muscles in the buttocks and shortens your muscles in the groin. Take a stand-up break for a minute at least once an hour. While standing lean back from the waist so as to stretch the muscles in the groin. Do this to the extent of comfort.

Do not sit with your legs under your chair. This causes the hamstrings to shorten, throwing your pelvis out of balance.

Do not sit with your legs crossed in the same manner all the time. Change which leg you put on top. When you cross your legs over top of one another you stretch the gluteus muscle on the side that is on top. After years of doing this you have succeeded in stretching the one side creating a muscle imbalance, which will destabilize your pelvis and you will become half-assed.

Remember all the advice about lifting with your knees bent? Do it!

As you start to exercise, start with core stabilizers.

Always work with a therapist who is familiar with chronic, not acute, back pain. The wrong therapy will cause you lots of agony.

CARPAL TUNNEL SYNDROME

Carpal tunnel syndrome involves pain in the thumb, index finger and half of the middle finger. Tendons of the wrist flexors pressing on the median nerve as they all go through the carpal tunnel cause the pain.

Surgery designed to open or widen the carpal tunnel is the most commonly used form of treatment. This treatment is almost 100% effective in reducing the pain in the short term. Long-term outcome studies are not as conclusive. One study reported that after one year the pain had returned in 98% of the subjects. I was privileged to have researched this phenomenon with a neurologist from the US. Treatment of muscle problems involving the sternomastoids (muscles in the neck) eliminated the carpal tunnel pain and corresponded with the expected changes in nerve conduction. An explanation of this phenomenon is outside the scope of this book.

Stretches of the neck, forearm, wrist and fingers are all necessary. For the wrist take the affected hand and bend it up and backwards towards the elbow stretching the muscles in the bottom of the wrist. All other muscles need to be stretched but the wrist stretch is most important.

Observations show that most people with carpal tunnel like to turn their head in the direction opposite to the side the pain is on. Correct this by simply occasionally turning your head to the side that hurts and hold it for about ten seconds. Repeat six times, three times daily. It is important that when the head is rotated it is kept in a level position making sure the chin does not drop down towards the shoulder.

Pain from the scalenes (muscle in the side of the neck) mimics that of carpal tunnel. Get a trigger point therapist to check the scalenes and do the appropriate stretches if trigger points are found.

Numerous shoulder muscles also send pain down the arm into the hand. These patterns often get mistakenly diagnosed as carpal tunnel syndrome. Get your shoulders checked for trigger points by a physical or massage therapist.

If you are wearing a wrist splint it is recommended that you only wear it to bed. During the day you need to maintain some degree of muscle tone while at night you don't want the pain to interfere with your sleep.

If your doctor wants you to take anti-inflammatory medication, take it. It will help with the inflammation at the carpal tunnel.

SHOULDER, NECK AND MID BACK PAIN

One of the most common causes of neck and shoulder pain today is incorrect posture while using computers.

Incorrect positioning of the mouse can cause pain in the shoulders. Working with the mouse in a reaching position with the arm extended out in front of the body creates muscle tension in the upper trapezius creating pain in this muscle. This muscle will cause pain in the shoulder as well as headaches. The correct position of the arm is to have the elbow in tight to the ribs, with the elbow directly below the shoulder forming a 90-degree angle at the elbow.

Pain in the mid back and neck is caused by incorrect elevation of the computer screen. The screen should be at eye level without the head being tilled.

Pain in the shoulders can also be caused by stress. As we get stressed there is a tendency for the shoulders to elevate moving closer to the ears. Take mini relaxation breaks every hour.

Shoulder pain can also come from incorrect workouts in which the muscles in the front of the chest (the pectoralis) become over developed as compared to the infraspinatus (the muscles that run over top of the shoulder blades). This will pull the shoulder forward out of neutral making it more susceptible to dislocation. I have seen this often when I trained a junior B hockey team.

Repetitive strain injury can happen to virtually any muscle in the body but in my experience the shoulder is the muscle most commonly affected. These injuries occur when a muscle is continually put under load due to incorrect posture, incorrect biomechanics or if the movement(s) continues beyond the muscle's ability to tolerate the load. As previously mentioned, if the mouse is positioned such that it positions the body too forward this puts a load on the upper trapezius. To combat this, stretch the affected muscle but also try using the other hand (usually left) to do some movement. This is awkward to learn but serves very quickly to balance the activity in the shoulders. Known as mirroring this technique has a sound neurological basis and can be utilized for any muscle in the body. I highly recommend this technique.

LEG AND FOOT PAIN

At the clinic two types of leg pain are seen most: a) pain that runs all the way down the side or back of the leg and b) pain in the ankle and foot.

Most of the people seen with pain down the leg have been diagnosed with sciatica. However pain from the gluteus minimus trigger points also goes down the leg. The health care professional often mixes these pain patterns up. True sciatic pain runs down the side of the leg and wraps around to the front of the knee. Pain from gluteus minimus trigger points run straight down the side of the leg or straight down the back of the leg never wrapping around to the front.

The gluteus minimus muscles are located on either side of the hip and serve to stabilize them during single limb weight bearing such as taking a stride. Often these trigger point disregulations develop in long distance runners and athletes that play on a smooth hard surface (i.e. asphalt). As these muscles are located deep in the hips, topical treatments often do not work. Stretching, myofascial release and muscle relaxants often work best. These muscles often compensate for problems in the Sacroiliac (SI) joints. These joints are found between the sacrum and the pelvis (see diagram below). Therefore, gait analysis, under the direction of a physical therapist, is also recommended.

While Dr. D has not had a great deal of experience in treating foot pain, the ones he has seen usually have pain in the ball or arch of the foot, which is referred from trigger points in the back of the calf. Stretching and icing these muscles usually takes care of these problems.

It is important to go for a walk when you have leg pain, as it helps with blood circulation. The more you sit the poorer your venous return becomes possibly leading to further complications. Personally I have found that even two minutes an hour helps clear the brain and reduce the swelling in the legs.

Another aide for leg pain and its associated swelling are elastic stockings that can cover the whole leg or just a part of it. These stockings fit tightly around the leg forcing the blood up into the body's core, thus reducing the swelling and pain. These were found to be very helpful especially post surgery.

To succeed in life, you need three things: a wishbone,
a backbone and a funny bone.

~ Reba McEntire ~

CHAPTER THIRTEEN

What is Chronic Benign Pain?

~~~~~~~

THROUGHOUT THE YEARS I HAVE BEEN PRIVILIDGED to see over a thousand chronic benign pain patients. The stories that follow are stories of actual patients seen at the clinic. They are intended to be representative of the type of patient seen and to illustrate the different problems and types of pain people bring to this office.

The first patient (Patient #1) is a 55-year-old woman who was referred by her friend. She has had 25 years of headache, a constant nagging headache at the left side (temple) of her head. Approximately once every four to six weeks she gets "a migraine," also on the left side, which incapacitates her for 2 – 3 days. She reports that she has to go to bed, cut all lights out, take medication and be quiet until she throws up. She takes 25 mgs of Elavil daily and Maxalt 5mgs as needed. Back pain and sinus problems are noted, but there are no other complaints. She is a senior executive in a major local company and also runs a ranch. She believes the headaches severely limit her abilities to move up in the company, plus she worries that she is neglecting her horses when down with a migraine.

Her history shows several falls from horses over the years with three or four concussions suspected. She presents as alert, orientated, and appropriate in affect with a little bit of evident anxiety. Numerous physicians and neurologists have been seen with tumors, etc. ruled out.

The next person to be seen (Patient #2) is a 64-year-old man who was injured in a motor vehicle accident five years prior. He was the driver of a car that was hit from behind by a speeding car. Upon presentation at the clinic, he complained of headaches (soreness bilaterally in the temples and on top of head), sore neck, left sided TMJ pain, aching in the right arm, problems with immediate memory and concentration, easily distracted, irritable, and anxious. He indicated that none of the symptoms had existed prior to the accident and were now varying in intensity from day to day. The pain had impacted his life such that he had retired from his occupation (plumber) and was quite desperate for help.

The last of my intakes for the day (Patient #3) is a 25-year-old woman who was referred to us with a diagnosis of fibromyalgia. She presented with pain in various parts of her body (neck, shoulders and right hip), fatigue, some slight sleep problems and depression with panic attacks.

The pain and panic attacks had been present since about age 15, however the sleep, fatigue and depression symptoms are recent developments. The pain is constant and gets worse when the panic attacks occur. She did not report any clearly-defined onset or precipitating event that caused the pain or started the panic attacks.

These cases are typical for our clinic. There is no consistent pattern that emerges from the intakes. The ages are varied (25 – 65), the length of time in pain varies from five years to 25, the type of pain and location varies, the intensity of the pain varies from constant to intense, the onset differs amongst them, and there is no one person that presents without emotional issues. As I sit here I am still struggling to figure out what chronic benign pain is. The list that follows outlines what I believe we know:

1. Chronic benign pain differs from acute pain, which involves different neurological systems. The neuroplasticity model makes the most sense to me. It explains how a system can adapt to repeated stimulation much in the same manner in which we learn to do things.

2. Chronic benign pain is defined by the length of time a person has it. It can start as early as six months and last for years.

3. The development of chronic benign pain is an insidious process in most cases. When there is a traumatic onset emotional factors are prevalent.

4. Chronic benign pain can be variable or constant or some combination thereof. It can severely hurt one day and be quiet the next.

5. Chronic benign pain saps your energy.

6. Chronic benign pain changes your life.

7. Chronic benign pain disrupts your sleep.

8. Chronic benign pain changes your personality. Your pre pain psychological status worsened by chronic pain.

9. Chronic benign pain demands your attention and clogs your brain.

10. The brain has a major role in chronic benign pain, yet sadly, is neglected by health care professions.

11. Muscles have a major role to play in chronic benign pain, yet are sadly neglected by health care professionals.

12. Chronic benign pain won't kill you, but it sure can wreck your life if you let it do so.

13. Chronic benign pain destroys hope and in turn is defeated by hope.

PATIENT #1

1 Given that headaches can come from multiple sources, including the neck, concussions and stress, a complete evaluation of patient #1was ordered. This included a qEEG, a trigger point evaluation, a SEMG examination and a psychophysiological stress test. This stress test utilizes eight different physiological measures to monitor an individual's physiology as they work under both stress and relaxation conditions. The eight measures include:

a) Heart rate

b) Heart rate variability

c) Respiration

d) Peripheral blood volume

e) Galvanic skin response

f) Peripheral temperature

g) Muscle tension

h) Brain wave activity

The results came back positive for muscle dysfunctions in the neck primarily involving the sternomastoids (SCMs) and cervical paraspinals (CPS). The pattern of muscle dysfunction was consistent with a whiplash injury which the patient recalled occurring when she was breaking a horse 26 years ago. Our assessment revealed that the whiplash likely created trigger points in the SCMs and CPS. These trigger points are identified as causing headaches. Stress showed up in the muscular system as well, compounding the problem.

After six weeks of neuromuscular retraining to fix the muscles in the neck and training for relaxation the patient reported no headaches. She has remained symptom free for fifteen months at the time of publication.

PATIENT #2

2 This gentleman had been in treatment with us for four months with some improvement such as a decrease in pain and an improvement in mood. However he was not improving as quickly as I had hoped, so we did a stress test. As his heart rate was found to be 120 beats per minute he was immediately sent to his doctor. His cardiologist suggested that his heart had been bruised by the shoulder harness but that otherwise he was in good shape. We worked on decreasing his stress and his heart rate with biofeedback.

We performed heart rate variability training, a form of biofeedback that helps reduce heart rate and overall arousal. After several months of treatment he dropped out to retire and enjoy life.

PATIENT #3

3 While this woman was diagnosed as having fibromyalgia, my opinion was that she was suffering from a number of myofascial pain syndromes that caused a depletion of her serotonin levels causing the panic attacks. During treatment, which, at print, is still ongoing, she has shown some improvement. The pain in her hip is now eliminated, but the other muscle pain remains. EEG neurofeedback helped with her depression, as have the panic attacks. Treatment has been ongoing for a year as she lives in another province and can only afford to come to the clinic occasionally – approximately once every six months. It is hoped that she can afford more treatment as she has responded well to what has been done so far. However if the problems in her neck and shoulders remain, it is feared that she will relapse, as these pain pathways will continue to be active and overload the central nervous system as previously noted.

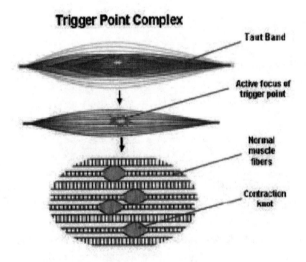

Trigger Point Complex of Myofascial Pain

So what is chronic benign pain? I don't believe there is a simple answer to this question. Chronic pain involves biomechanical factors, biochemical factors, as well as central, peripheral, and autonomic nervous system factors, emotional factors, genetic and psychosocial factors, all interacting in an unique manner for each individual: a complex problem with no simple answers.

ENJOY WHEN YOU CAN, AND ENDURE WHEN YOU MUST.

~ JOHANN WOLFGANG VON GOE ~

CHAPTER FOURTEEN

How Did I Develop Chronic Pain?

~~~~~~~

AS I SIT HERE WORKING ON THIS MATERIAL I have virtually no pain. Yesterday I was grouchy, bitchy and hurt all over. Some days I am really good in which pain is minimal and other days I can barely walk and stand. When the pain is at its worst it is located primarily in my hips and knees with the left side generally hurting more than the right. Occasionally I get low back pain as well.

So how did I develop chronic pain from a simple slip and fall? The pain initiated from the fractures and dislocation. This is what I call the causal factors (such as the fall from a horse, or motor vehicle accident as noted in the examples above).

As the sensory component from the fractures and dislocation decreased other factors became involved. These are the factors which I call maintenance factors. I believe causal and maintenance factors are different from each other. Causal factors:

1.   My illnesses right from birth set my nervous system up to be sensitive to pain signals.

2.   The Leg Perthes Syndrome produced a biomechanical imbalance in my pelvis that was not corrected for 55 years.

3.   My biomechanical imbalance and polio altered my motor control and muscle functioning, having to adapt to the leg length difference.

4.   The polio may have created a biochemical problem at the neuromuscular junctions creating a post polio syndrome.

5.   My left and right muscles were not of equal length, and consequently not of equal strength.

6.   The fall compounded muscle issues by displacing the head of the femur (thigh bone) leaving the ligaments and muscles slightly out of previous alignment.

7.   The force of the fall was transmitted up my spine stopping in my brain, clouding it up.

8.   Psychologically, my predisposition to depression precipitated my withdrawal from my support group, which led to more depression and withdrawal.

9.   Over the years my pain was specific to different sites leading me to focus my attention on those individual sites and ignoring the big picture. I believe this was an absolutely critical factor in maintaining my pain. When my focus was on the specific sites I would behave in a manner designed to avoid more pain and possible injury. After the second surgery when the pain was less but not gone I finally realized I was suffering from chronic benign pain. My attitude and behaviors changed to start challenging the pain versus avoiding it.

10.   Also very importantly I did not like the effect drugs had on me so I did not tend to use them, thus avoiding their compounding of my pain. Drugs can cause something called iatrogenic pain: pain caused by inappropriate or incorrect medical procedures. This is a small but very powerful part of the chronic pain problem.

       In conclusion it appears to me that there were a number of factors that contributed to the maintenance of the pain. How a medication is expected to solve these issues is beyond me.

       It has now been 11 years during which I have certainly been taught a few lessons. As a person I hope that I am much more humble than I was previously, and more aware of others' needs. I do believe I am a better therapist than I was before, as I understand chronic pain patients better. This understanding allows me to assess the pain situation more accurately. I hope I have become more tolerant of those who are genuinely in pain and non supportive of those who wallow in self-pity and grief. Throughout all of this I continue to look at the involvement of muscles in chronic pain and gradually focus on the brain as an active part of the chronic pain cycle.

WE MAKE A LIVING BY WHAT WE GET; WE MAKE A LIFE BY WHAT WE GIVE.

~ WINSTON CHURCHILL ~

CHAPTER FIFTEEN

The Conclusion

~~~~~~~

IT HAS BEEN 11 YEARS since I started on this epic voyage, learning a lot. I hope I have become a better person, and a better therapist.

In looking back, several things have become apparent to me:

Drugs are not the answer. Just purely and simply, drugs will not cure chronic pain. They will mask it, and unfortunately can even make it worse.

Knowledge concerning muscle activity and myotatic units is not that widely spread, particularly as it relates to chronic pain. Forty percent of the human body is comprised of muscle. Thus programs that do not include muscles in their investigations are severely handicapped in arriving at correct diagnoses.

Almost all chronic pain programs do not pay any attention to the brain. How science can disregard this organ absolutely blows my mind.

The health care system is designed to treat acute pain. It does a reasonable job of this. It is designed for surgery, drugs and rest and does not work for chronic pain. Medical science still does not know what causes chronic pain and continues to apply acute pain care methodology to this problem. I was able to steer through many of the potential iatrogenic problems mainly by ignoring some of the recommendations from well-meaning therapists.

One thing more than anything sticks out at me. Chronic pain at the present time is not "curable" in the strictest sense of the word. **People with chronic pain will not return to the state they were in before the pain began. However they can learn to significantly reduce the pain, how the pain affects their life, and most importantly how they can take back control of their life.**

It is this latter point that I wish to really emphasize. Chronic pain patients have to make a decision: are they going to control the pain, or are they going to let the pain control them? This book, this story, has been about how I learned to reduce the pain and get on with life. I still have pain, although it is quite variable now. I can't walk without a limp, can't walk extended distances, can't climb stairs with something in my hands, and can't square dance. I couldn't before anyway. But I also think of all the things I can do. I can cut the lawn, go for walks with my dogs, attend Rotary meetings and see friends, be a partner to Mary, and do all sorts of other things. The list of what I can do now far exceeds the list of things I can't.

I have become a better therapist. I listen more closely to my patients, understanding the frustrations and anger they experience, and I'm more demanding that they lose the helpless, hopeless role. More so, I understand the way the health care system work, or doesn't, as the case may be, and how it affects the chronic pain patient.

I also understand the health care insurance system. The chronic pain patient often winds up feeling or being abandoned by the system, the medical practitioner, the insurance agent and everyone else in between.

It is no wonder that alternative health care is so promising. Yet this alternative is blocked by well-meaning government legislation and politics. Critics of the current medical system have stated that it is becoming increasingly evident that politics and not science plays a role in what products are accepted and which ones are refused a license.

So at the end of the day there remains one person the chronic pain patient can count on: him or herself. It is hope that carries the chronic pain patient through. Start using that to change your life and take charge of it. Monitor your behaviors that cause pain and start to change them, one at a time, not all at once. Start exercising, start doing mental workouts and start getting out to become less socially isolated. Use the materials in this book and materials from the books listed in the appendix to work towards pain reduction. Work with your doctor and any other health care provider you use. Have knowledge of chronic pain issues, move forward into better health, for you can do it.

I will leave you with a prayer for health and happiness and with a thought from Mark Twain:

> "If you always do what you have always done,
> then you'll always get what you always got."

ONE SHIP SAILS EAST,
AND ANOTHER WEST,
BY THE SELF-SAME WINDS THAT BLOW,
TIS THE SET OF THE SAILS
AND NOT THE GALES,
THAT TELLS THE WAY WE GO.

~ ELLA WHEELER WILCOX ~

EPILOGUE

~~~~~~~

My hands are sweaty, my heart is racing, and my stomach feels like its playing handball against my intestines. It almost feels like a panic attack except my mind is crystal clear, sharp as a razor.

It is 10 minutes to eight a.m., March 10, 2011 in New Orleans. At eight a.m. I am scheduled to start speaking about chronic pain to a group of my International colleagues. I don't know what to expect — not of them, but of me.

As Dr. Stuart Donaldson, I have lectured all over the world on pain related topics. I have won numerous accolades, been recognized for my research, and won International Awards. Yet here I am, scared skinny. However modest my successes in the past, I am afraid today's performance will be compared to those from the past and found sadly lacking. I am scared that at the end of the day they will say, "Boy, he has certainly lost it."

You see it has been five years since I have presented papers or workshops. I have no confidence and no idea how I will do. Five years lost to chronic pain, leaving me not knowing if I can perform and deliver at the same level I have in the past. I don't remember the jokes I used to tell or demonstrations involving the audience I previously used.

More worrisome, I am having slight brain fog (post surgery due to the anaesthetic), having trouble with word searches (putting the name to an object). I hope I don't make a complete fool of myself, or God forbid, people think I don't know what I am talking about.

It is eight o'clock, show time. I walk into the room and say "Good Morning Ladies and Gentlemen and anyone else in the room. My name is Stuart Donaldson and we are here today to talk about chronic pain."

I say to the group, "This lecture is very dear to me as I have spent years teaching and studying chronic pain and more recently, years living with it. My hope at the end of the day is that my professional and personal experiences from sitting on both sides of the desk will contribute to your greater understanding of this thing we call chronic pain. I hope it will make you a better therapist. I hope you will be able to help your patients to an even greater degree than you do now."

As I stood there in front of everyone I think how amazing it is that people would pay good money to come and hear me speak. I always thought this even before the development of the pain. They could be out drinking, eating, and enjoying themselves, but here they are, by choice, to listen to me.

Professionally, on one side of the desk sits Dr. Stuart Donaldson. I have a PhD in Clinical Psychology with postdoctoral experience in rehabilitation psychology, studying at the University of Calgary and graduating in 1989. Presently I am an Adjunct Professor in Applied Psychology at the University. Dr. D, as I am known and like to be called, has been privileged to be trained in biofeedback procedures and has used these techniques to treat hundreds of chronic pain patients, usually decreasing their pain and publishing the results.

Over the years, as the clinic's reputation grew and success in treatment grew, I became more and more confident in my abilities to treat this difficult condition known as chronic pain. My initial focus was on low back pain, particularly myofascial pain and how to successfully treat it with biofeedback procedures using SEMG techniques. I became successful and well known, lecturing on these techniques throughout the world, culminating in 1995 in winning the prestigious award "The Outstanding Contribution To The Interdisciplinary Pain Management" literature as presented by the *American Journal of Pain Management*.

In 1995 the clinic Myosymmetries added EEG neurotherapy to their treatment procedures leading us into the world of fibromyalgia and issues involving the brain. Again the clinic became reasonably successful in treating and researching these dysfunctions, again leading to lectures and numerous trips. Throughout this time my work(s) through publication in scientific journals became well known and I was privileged to work with some of the foremost authorities in the world in the myofascial, fibromyalgia and chronic pain fields.

In 2000 the National Fibromyalgia Association and the Government of Canada appointed me to the Expert Consensus Panel to study and develop a consensus document about fibromyalgia. I remember meeting the other members and being slightly overwhelmed, as I was the only PhD on the panel. The rest were medical doctors — some of whose works had greatly impressed me. But by then I had some confidence, thinking I was reasonably good in treating chronic pain.

Back at the lecture in New Orleans Stu, the patient on the other side of the desk, begins to appear. As I stand/sit/lean and move around while I lecture I am aware of numerous pain sensations that occur individually and together. Both legs and feet go numb, sharp pain runs down the outside of my left leg between the hip and knee, my bum hurts, then my back. I am constantly shifting as this relieves the pain temporarily until it is replaced by something else. I lack stamina and have to change my lecture style to get more group discussion, so I can take a mental break. I had to get a chair that looks more like a bar stool brought into the room, so I could sit, stand and lean as needed.

As the personal side of the lecture unfolds I become aware of several different feelings and emotions: anger, fear, sadness, grief, happiness, elation, all combined in a ménage. This is the part of the lecture that won't be found in textbooks to any great degree. I believe this adds much value and insight to the lecture, for there is no greater teacher than experience.

My world changed on that New Year's Eve in 2000 when I slipped and fell in front of a restaurant. Since then I have been in constant pain.

Into the professional world of Dr. D came the personal world of Stu. I expected that as a recognized pain expert, I would be able to treat myself and get on with life as it existed before the accident. What a shock when that didn't occur. It changed my life and family forever. My realization that the things I knew professionally weren't working was not abrupt. It was a gradual realization that was coupled with anger and a lot of denial.

These experiences, the ones on the other side of the desk, were what I shared with this audience of health care professionals, and somehow it worked! My talk was well received and I continue to work the speaker's circuit, as well as consult with anyone who may find my insights useful to the treatment of a patient of theirs.

Writing this book has been a gift for me. Expressing the thoughts, feelings, frustrations and limitations around the pain that is in my life has been cathartic, but beyond that, it has been a journey to the other side of the desk. It is my sincere wish that this volume is somehow helpful to others, and in that way, it becomes also a gift for you.

HEALTH IS THE GREATEST GIFT, CONTENTMENT THE GREATEST WEALTH, FAITHFULNESS THE BEST RELATIONSHIP.

~ BUDDHA ~

GLOSSARY

~~~~~~~

Acetabulum: A deep socket on the outer surface of the pelvis (hip bone) into which the head of the femur (thigh bone) fits forming the hip joint. (Wikipedia)

AP views: refers to an anterior – posterior view in an X-Ray in which the bone is viewed from the front of the patient towards the back.

Biochemicals: chemicals that are manufactured naturally in the body.

Biofeedback: Biofeedback is the process of becoming aware of various physiological functions using instruments that provide information on the activity of those same systems, with a goal of being able to manipulate them at will. (Wikipedia)

Cognitive behavioral therapy: Cognitive behavioral therapy is a psychotherapeutic approach that aims to solve problems concerning dysfunctional emotions, behaviors and cognitions through a goal-oriented, systematic procedure. The title is used in diverse ways to designate behavior therapy, cognitive therapy, and to refer to therapy based upon a combination of basic behavioral and cognitive research. (Wikipedia)

Dorsal Horn: of the spinal cord is the dorsal (more towards the back) grey matter of the spinal cord. It receives several types of sensory information from the body, including light touch, proprioception, and vibration. (Wikipedia)

EEG neurotherapy/biofeedback: This is a specific biofeedback process in which the activity of the brain (brain waves) is provided (displayed) to an individual for the purpose of changing the brain waves at will.

EMDR (Eye movement desensitization and reprocessing): is a comprehensive treatment approach that integrates elements of effective psychodynamic, imaginal exposure, cognitive therapy, interpersonal, experiential, physiological and somatic therapies. It also uses the unique element of bilateral stimulation (e.g. eye movements, tones, or tapping (Wikipedia)

Gate Control Theory of Pain: suggests that there is a "gating system" in the central nervous system that opens and closes to let pain messages through to the brain or to block them. According to the gate control theory of pain, our thoughts, beliefs, and emotions may affect how much pain we feel from a given physical sensation.

Glasgow Coma Scale: is a neurological scale that aims to give a reliable, objective way of recording the conscious state of a person

Iliocostalis lumborum: deep back (erector spinae) muscle; origin, posterior aspect of sacrum and thoracolumbar fascia; insertion, the angles of lower six ribs; action, extends, abducts, and rotates lumbar vertebrae; (Wikipedia)

Infraspinatus: is a thick triangular muscle, which occupies the chief part of the infraspinatus fossa. The infraspinatus is a muscle of the rotator cuff (Wikipedia)

Legg–Calvé–Perthes syndrome: a degenerative disease of the hip joint, where growth/ loss of bone mass leads to some degree of collapse of the hip joint and to deformity of the ball of the femur and the surface of the hip socket. The disease is typically found in young children, and it can lead to osteoarthritis in adults. (Wikipedia)

Myofascial Pain: a condition characterized by acute and chronic pain both localized and referred. It is associated with and caused by "trigger points" (TrPs), which are localized and sometimes extremely painful contractures ("knots") found in any skeletal muscle of the body. Each muscle refers pain in its own unique manner.

Motor-neuron junctions: applies to neurons located in the central nervous system (or CNS) that project their axons outside the CNS and directly or indirectly control muscles.

Motor Homunculus: a term used to describe the distorted scale model of a human drawn or sculpted to reflect the relative space human body parts occupy on the somatosensory cortex (sensory homunculus) and the motor cortex (motor homunculus).

Myotatic units: A group of muscles that function together as a unit because they share common spinal-reflex responses.

Neuroplasticity: is the changing of neurons, the organization of their networks, and their function via new experiences.

Neuro-transmitters: are endogenous chemicals which transmit signals from a neuron to a target cell across the synapse

Pathophysiologies: the physiology of abnormal states; *specifically*: the functional changes that accompany a particular syndrome or disease. (Wikipedia)

Pectoralis: is a thick, fan-shaped muscle, situated at the chest (anterior) of the body. It makes up the bulk of the chest muscles in the male and lies under the breast in the female

Psychophysiology: is the branch of psychology that is concerned with the physiological bases of psychological processes.

Quantitative electroencephalography (qEEG): is a technique in which EEG data is subjected to Fast Fourier analysis and then compared to a data base to determine the absolute power of various frequencies found in the brain.

Rheumatologists: is a sub-specialty in internal medicine and pediatrics, devoted to the diagnosis and therapy of conditions and diseases affecting the joints, muscles, and bones.

Scalenes: are a group of three pairs of muscles in the lateral (either side) of the neck.

Sciatica: is a set of symptoms including pain that may be caused by general compression and/or irritation of one of five spinal nerve roots that give rise to each sciatic nerve,

Serotonin: is a monoamine neurotransmitter, biochemically derived from tryptophan, that is primarily found in the gastrointestinal (GI) tract, platelets, and central nervous system (CNS) of humans and animals

Surface electromyographic (SEMG): SEMG refers to a technique in which electrodes are placed on the skin over a muscle for purposes of studying that muscles activity in relation to other muscles.

Sternomastoids: a paired muscle in the superficial layers of the anterior portion of the neck. It acts to flex and rotate the head.

Trapezius: the trapezius is a large superficial muscle that extends longitudinally from the occipital bone to the lower thoracic vertebrae and laterally to the spine of the scapula (shoulder blade). Its functions are to move the scapulae and support the arm.

SUGGESTED READING

~~~~~~~

(in no particular order)
These are books that Dr. D has read over the years,
which will hopefully help you.

Brain

Amen, D. *Change Your Brain Change Your Life*. Three Rivers Press

Amen, D. *Making the Good Brain Great*. Three Rivers Press

Amen, D. *Change Your Brain Change Your Body*. Three Rivers Press

Doidge, N. *The Brain That Changes Itself*. Penguin Books

Katz, L C. & Rubin M. *Keep Your Brain Alive*. Workman Publishing.

Robbins, J. *A Symphony in the Brain*. Atlantic Monthly Publishing

Sears W. & Thompson, L. *The A.D.D. Book*. Little, Brown & Company

Taylor, Jill B. *My Stroke of Insight*. Penguin Group (USA). New York

Muscles

Basmajian, J. & De Luca, C. *Muscles Alive: Their Functions Revealed by Electromyography*. Fifth Edition. William & Wilkins.

Cailliet, R. *Soft Tissue Pain and Disability*. 2nd Edition. F.A. Davis Company.

Travell J. & Simons, D. *Myofascial Pain and Dysfunction: The Trigger Point Manual: The Upper Extremities. Volume 1*. William & Wilkins.

Travell J. & Simons, D. P*ain and Dysfunction: The Trigger Point Manual: The Lower Extremities. Volume 2*. William & Wilkins.

Sella, G. *Muscles In Motion: The SEMG of the ROM of the Human Body*. GENMED Publishing, Martins Ferry, OH.

Sella, G **"Guidelines for Neuro-muscular Re-education with S-EMG/ Biofeedback"** GENMED Publishing, Martins Ferry, OH.

Stretching

Anderson, B. *Stretching Revised Edition*. Shelter Publications

Stark, S. D., *The Stark Reality of Stretching*. Stark Reality Corp.

Chronic Pain

Hendler, N., Long D. & Wise T. *Diagnosis & Treatment of Chronic Pain*. John Wright * PSG Inc

Kleinman, Arthur, *The Illness Narratives: Suffering, Healing, and the Human Condition*. Basic Books. New York.

Koestler. Angela J, Myers, Ann. *Understanding Chronic Pai*n. University Press of Mississippi: Jackson, MS

Ostalecki, S. *Fibromyalgia: The Complete Guide from Medical Experts and Patients*. Jones & Bartlett Publishers.

Turk, D, Meichenbaum, D & Genest,M. *Pain & Behavioral Medicine: A Cognitive Behavioral Approach*. Guilford Press.

Turk, D. & Winter F. *The Pain Survival Guide*. American Psychological Association.

Emotions

Burns, D. *The Feeling Good Handbook*. Plume Publications.

Otis, John D. *Managing Chronic Pain: A Cognitive Behavioral Therapy Approach*. Oxford Press. New York.

Posen, D. *The Little Book of Stress Relief.* Key Porter Books.

CPSIA information can be obtained at www.ICGtesting.com
Printed in the USA
LVOW11s1330080913

351488LV00001B/6/P